AF737331

Caribbean Geography

A Scholarly Bibliography

Thomas A. Rumney

THE SCARECROW PRESS, INC.
Lanham • Toronto • Plymouth, UK
2012

Published by Scarecrow Press, Inc.
A wholly owned subsidiary of
The Rowman & Littlefield Publishing Group, Inc.
4501 Forbes Boulevard, Suite 200, Lanham, Maryland 20706
www.rowman.com

10 Thornbury Road, Plymouth PL6 7PP, United Kingdom

Copyright © 2012 by Thomas A. Rumney

All rights reserved. No part of this book may be reproduced in any form or by any electronic or mechanical means, including information storage and retrieval systems, without written permission from the publisher, except by a reviewer who may quote passages in a review.

British Library Cataloguing in Publication Information Available

Library of Congress Cataloging-in-Publication Data
Rumney, Thomas A.
Caribbean geography : a scholarly bibliography / Thomas A. Rumney.
p. cm.
Includes bibliographical references and index.
ISBN 978-0-8108-8303-1 (cloth : alk. paper) — ISBN 978-0-8108-8304-8 (ebook)
1. Caribbean Area—Geography—Bibliography. I. Title.
Z1595.R86 2012
[F2158]
015.729—dc23 2012004979

™ The paper used in this publication meets the minimum requirements of American National Standard for Information Sciences—Permanence of Paper for Printed Library Materials, ANSI/NISO Z39.48-1992. Printed in the United States of America.

CONTENTS

INTRODUCTION

The islands and seascapes gracing the Caribbean Sea have long been areas of interest and research for geographers and other scholars from around the world. More than beautiful beaches and "fun places in the sun," though certainly they are just that, the lands and waters of the Caribbean region have stimulated an extensive body of research and writing across the many fields of geography. It is the intent of this compendium to collect, organize, and present as many of these scholarly publications as possible to aid in the teaching, study, and further scholarship of the geography of this area.

This geographical study of the Caribbean region, of necessity, begins with an identification of the lands and countries of the area. By far, the largest parcels of land in the area are the Greater Antilles Islands. These include Cuba, Hispanola (divided into the countries of Haiti and the Dominican Republic), Jamaica, and the Commonwealth of Puerto Rico. The second large collection of islands traditionally has been called the Lesser Antilles, mainly due to their relative sizes. These islands include the British and American Virgin Islands, Anguilla, Barbuda, Antigua, St. Kitts, Nevis, Montserrat, Guadeloupe, Dominica, Martinique, St. Vincent and the Grenadines, St. Lucia, Barbados, Grenada, Trinidad and Tobago, Aruba, Bonaire, and Curacao. Plus, although not usually considered parts of the Antilles, other nearly islands are included as well in terms of publications. They are Bermuda, the Bahamas, the Turks and Caicos, and the Caymans.

Caribbean Geography: A Scholarly Bibliography has seven parts: General Works, Cultural and Social Geography, Economic Geography, Environmental and Physical Geography, Historical Geography, Political Geography, and Urban Geography. The types of publications noted include atlases, books, chapters of books, articles from scholarly journals, master's theses, and doctoral dissertations. Most of these are written in English, but there are also works written in French, German, Spanish, Dutch, and other languages. These often have been translated into English, and the original languages are noted within these citations. Entries in each chapter are arranged alphabetically by author's last name. Where there is more than one publication per author, the earliest is listed first and the rest are listed chronologically.

While the totality of entries listed here represents probably the vast majority of research and published works done by geographers on the various aspects of the geographical study of the Caribbean region, it is also probable that something has been missed. This is the responsibility of the author and not the publisher. It is hoped that two major types of services are provided for the users of this volume. The first, and most obvious, is that this is a convenient and useful collection of existing references on the geography of the Caribbean region that can assist teachers and students in both the study and research of the area. The second service is to provide an image of what has not been studied and published on the many varied aspects of the region's geographies. Such knowledge could encourage more efforts by geographers to further our knowledge and understanding of the broader Caribbean region's dynamic geography. In addition to the included books, atlases, theses, and chapters, the following scholarly journals have served as sources of entries for this compendium.

Ambio
American Journal of Economics and Sociology
Annales de Geographie
Annals, New York Academy of Sciences
Annals of the Association of American Geographers
Annals of Tourism Research
Antipode
Applied Geography
Applied Geography and Development
Applied Geography Studies
Area
Belgeo
Biotropica
Bolletino della Societa Geografica Italiana
Bulletin, Illinois Geographical Society
Bulletin de l'Association de Geographes Francais

Bulletin of the Association of North Dakota Geographers
Bulletin of the Philadelphia Geographical Society
Les Cahiers d'Outre Mer
Canadian Geographer
Canadian Journal of Development Studies
Caribbean Geography
Caribbean Journal of Science
Caribbean Quarterly
Caribbean Studies
Cities
Climatological Bulletin
Coastal Management
Contemporary Review
Earth Surface Processes and Landforms
Economic Development and Cultural Change
Economic Geography
Ecumene (first version)
Ecumene/Cultural Geographies
Ekistics
Environment and Planning A
Environmental Conservation
Environmental Management
Die Erde
Erdkunde
Fennia
Florida Geographer
Focus
Forest and Conservation History
Geoforum
Geografiska Annaler B
Geographia Polonica
Geographical Bulletin
Geographical Journal
Geographical Magazine
Geographical Perspective
Geographical Review
Geographical Survey
Geographie et cultures
Geographische Zeitschrift
Geography

Geography Compass
GeoJournal
Historical Geography
The Holocene
Human Ecology
Hydrological Journal
Interamerican Economic Affairs
International Journal of Heritage Studies
International Journal of Tourism and Hospitality Research
International Journal of Tourism Management
International Migration Review
Island Studies Journal
Jamaican Historical Review
Jamaica Journal
Jamaican Naturalist
Journal of Biogeography
Journal of Borderland Studies
Journal of Caribbean Science
Journal of Coastal Research
Journal of Cultural Geography
Journal of Developing Areas
Journal of Eastern Caribbean Studies
Journal of Geography
Journal of Historical Geography
Journal of Hydrology
Journal of Latin American Geography
Journal of Rural Studies
Journal of the Barbados Museum and Historical Society
Journal of Travel and Tourism Marketing
Journal of Tropical Geography
Leisure Sciences
Middle States Geographer
Monthly Weather Review
Mountain Research and Development
National Geographical Journal of India
Natural Hazards
Ohio Geographer
Operational Geographer
Pennsylvania Geographer
Philippine Geographical Journal

Physical Geography
Places
Political Geography Quarterly/Political Geography
Population, Space, and Place
Proceedings, Association of American Geographers
Proceedings, New England-St. Lawrence Valley Division, Association of American Geographers
Professional Geographer
Progress in Development Studies
Progress in Physical Geography
Quarterly Journal of the Royal Meteorological Society
Quaternary Research
Revista de Ciencias Sociales
Revista Geografica
Rocky Mountain Social Science Journal
Singapore Journal of Tropical Geography
Social and Cultural Geography
Social and Economic Studies
Social Science and Medicine
Southeastern Geographer
Soviet Geography
Terra Incognita
Third World Planning Review
Tijdschrift voor Economische en Sociale Geographie
Tourism
Tourism Geographies
Town and Country Planning
Transactions, Institute of British Geographers
Transactions, Pioneer America Society
Urban Geography
Urban Studies
Weather
Yearbook, Association of Pacific Coast Geographers
Yearbook, Conference of Latin Americanist Geographers
Zeitschrift fur Wirtschaftsgeographie

PART 1

General Works

Abrahams, Peter. *Jamaica: An Island Mosaic.* London, England: HMSO, 1957.

Aguayo, Alfredo M., and Carlos de la Torre y Huerta. *Tratado elemental de geografia de Cuba para des las escuelas.* Havana, Cuba: La moderna poesa, 1907.

Ahmad, Rafi. "Natural Hazard Maps in Jamaica." *Caribbean Geography,* 12, No. 2 (2001): 90-107.

Alexander, W.H. "St. Christopher, West Indies." *Bulletin of the American Geographical Society,* 33, No. 1 (1901): 42-46.

Anderson, Glenn. "Florida's Bahamian Connection." *Florida Geographer,* 27 (1996): 68-80.

Anglade, Georges. *Atlas critique d'Haiti.* Montreal, Quebec: Universite de Quebec de Montreal, 1982.

Antonini, Gustavo. "Process and Patterns of Landscape Change in the Linea Noroeste, Dominican Republic." Ph.D. Dissertation, Columbia University, New York, N.Y., 1968.

Ardouin, B. *Geographie de l'ile d'Haiti.* Port-au-Prince, Haiti: n.p., 1832.

Atlas de Cuba. Havana, Cuba: Instituto Cubano de Geodesia y Cartografia, 1978.

Atlas d'Haiti. Talence, France: Universite de Bordeaux, 1985.

Atlas for the Eastern Caribbean. London, England: Collins-Longman, 1974. 57 pp.

Atlas for Jamaica and the Western Caribbean. Glasgow, Scotland: Collins-Longman Atlases, 1971 and 1978. 57 pp.

Atlas National de Cuba. Havana, Cuba: Academia de Ciencias de Cuba, 1970.

Augelli, John P. "Geography of Agriculture and Settlement in Interior Puerto Rico: Problems of the Upper Loiza Valley." Ph.D. Dissertation, Harvard University, Cambridge, Mass., 1950.

Augelli, John P. "The British Virgin Islands: A West Indies Anomaly." *Geographical Review*, 46, No. 1 (1956): 43-58.

Augelli, John P. "The Dominican Republic." *Focus*, 10, No. 6 (1960).

Augelli, John P. "The Dominican Republic." *Focus*, 15, No. 9 (1965).

Augelli, John P., ed. *Caribbean Lands*. Grand Rapids, Mich.: Fideler, 1965.

Augelli, John P. *Puerto Rico*. Boston, Mass.: Ginn, 1973.

Babcock, W.H. "Atlantis and Antillia." *Geographical Review*, 3 (1917): 392-395.

Babcock, W.H. "Antillia and the Antilles." *Geographical Review*, 9 (1920): 109-124.

Backus, John G.H. "The Geography of Curacao." Master's Thesis, Columbia University, New York, N.Y., 1963.

Baetens, Eddy H. "Curacao: Restructuring in the 1980s and 1990s." *Caribbean Geography*, 3, No. 2 (1991): 126-131.

Bahamas Ministry of Archives. *Atlas of the Commonwealth of the Bahamas*. Kingston, Jamaica: Kingston Publishers, 1985.

Bahr, Jurgen, and Sonke Widderich. *Vom Nostrand zum Normalzustand: Eine Bilanz des Kubanischen Transformationsprozesses*. Kiel, Germany: Kieler Geographische Schriften, Band 103, 2000.

Balchin, W.G., and A. Coleman. "Puerto Rico." *Geography*, 50 (1965): 274-286.

Barker, David, David Dodman, and Duncan McGregor. "Caribbean Vulnerability and Global Change: Contemporary Perspectives." Pp. 3-21 in *Global Change and Caribbean Vulnerability: Environment, Economy, and Society at Risk*, edited by Duncan McGregor, David Dodman, and David Barker. Mona, Jamaica: University of West Indies Press, 2009.

Barker, David, Sally Lloyd-Evans, and Duncan McGregor, eds. *Sustainability and Development in the Caribbean: Geographical Perspectives*. Kingston, Jamaica: University of West Indies Press, 1998.

Barker, David, and Duncan McGregor, eds. *Environment and Development in the Caribbean: Geographical Perspectives*. Kingston, Jamaica: Press of the University of the West Indies, 1995.

Barker, David, and David J. Miller. "Hurricane Gilbert: Anthropomorphising a Natural Disaster." *Area*, 22 (1990): 107-116.

Barker, David, Carol Newby, and Michael Morrissey, eds. *A Reader in Caribbean Geography*. Kingston, Jamaica: Ian Randle, 1998.

Bastistide, J.C., et M. Etna. "Le Sud Basse-Terrien (Guadeloupe): approche geographiques." *Les Cahiers d'Outre Mer*, 36 (1983): 49-74.

Baume, O. et al. "Entwicklung und Stand der geographischen Landschaftsforschung in Kuba." *Petermanns Geographische Mitteilungen*, 138, No. 4 (1994): 235-244.

Baume, O. et al. "Okonomischen Landschaftsbewertung durch Potenzial bestimmung und aktuelle Immobilien preise als Basis einer Nachhaltigen Umwelst politik in der Provinz Ciudad de la Habana, Kuba." *Geo-oko*, 26, No. 1-2 (2005): 95-111.

Bell, Ion. *The Dominican Republic*. Boulder, Colo.: Westview, 1981. 392 pp.

Bent, R.M., and E. Bent-Golding. *A Complete Geography of Jamaica*. London, England: Collins, 1966. 128 pp.

Berringuier, C. "L'espace regional Martiniquais." *Les Cahiers d'Outre Mer*, 20 (1967): 150-184.

Billmyer, James. "The Cayman Islands." *Geographical Review*, 36, No. 1 (1946): 29-43.

Blakemore, Harold, and C.T. Smith, eds. *Latin America: Geographical Perspectives*, 2nd ed. London, England: Methuen, 1983. 555 pp.

Bloch, P. "Kleine Landeskunde der Dominikanischen Republik." *Geographische Rundschau*, 12, No. 9 (1960): 351-359.

Bloch, P. "Kuba 1963." *Geographische Rundschau*, 15 (1963): 291-297.

Blouet, Brian, and Olwyn M. Blouet. *Latin America: An Introductory Survey*. New York, N.Y.: Wiley, 1982. 300 pp.; 3rd ed., New York, N.Y.: John Wiley, 1997. 550 pp.

Blume, Helmut. "Die Inselwelt der Antillen." *Bild der Wissenschaft*, 10 (1970): 998-1005.

Blume, Helmut. *The Caribbean Islands*. London, England: Longman, 1974.

Boehm, Richard G., and Sent Visser. *Latin America: Case Studies*. Dubuque, Iowa: Kendall/Hunt, 1984. 300 pp.

Boswell, Thomas D., and James R. Curtis. *The Cuban-American Experience: Culture, Images, and Perspectives*. Totowa, N.J.: Rowan and Littlefield, 1984. 200 pp.

Bounds, John H. "The Bahamas." *Focus*, 19, No. 9 (1969).

Bristol, Charles L. "Notes on the Bermudas." *Bulletin of the American Geographical Society*, 33, No. 3 (1901): 242-248.

Brooke, M.E. "Casual Glimpses of St. Thomas, the Island of Black Beard and Bay Rum." *Bulletin of the Geographical Society of Philadelphia*, 27 (1929): 43-53.

Brothers, Timothy S., Jeffrey S. Wilson, and Owen P. Dwyer. *Caribbean Landscapes: An Interpretive Atlas*. Coconut Creek, Fla.: Caribbean Studies Press, 2008.

Brown, David. "The Geography of Antigua." Master's Thesis, University of Chicago, Chicago, Ill., 1949.

Burac, Maurice. *Les Petites Antilles : Etude geographique des disparites regionales de developpement*. Talence, France: Presses Universitaires de Bordeaux, 1989.

Bureau, Jacques. "Les traits dominants de la geographie de Sainte-Lucie (B.W.I.)." Master's Thesis, Universite Laval, Ste. Foy, Quebec, 1965.

Burstein, Harold N. "The British West Indies." Master's Thesis, George Washington University, Washington, D.C., 1963.

Bush, Davis M. et al. *Living with the Puerto Rico Shore*. Durham, N.C.: Duke University Press, 1995. 193 pp.

Canet, G. *Atlas de Cuba*. Cambridge, Mass.: Harvard University Press, 1949.

Capone, D.L., and A.F Ryan. "The Regional Sea: A Theoretical Division of the Gulf of Mexico and the Caribbean Sea." *Transactions, Miami Geographical Society*, 3 (1973): 1-9.

Chantada, Amparo. *La geografia en Santo Domingo*. Santo Domingo, R.D.: Editora Universitaria-UASD, 1987.

Chaunu, P. "Paradoxes porto-ricains." *Les Cahiers d'Outre Mer*, 3 (1950): 166-176.

Clarke, Colin G. *Jamaica in Maps: Graphic Perspectives of a Developing Country*. New York, N.Y.: Africana, 1974. 104 pp.

Clarke, Colin G. "Insularity and Identity in the Caribbean." *Geography*, 61, No. 1 (1976): 8-16.

Clarke, W.C. *Notes on the Geography and History of Dominica*. Berkeley, Calif.: University of California, Department of Geography, 1962.

Clay, Roger C., and L. Alan Eyre. *The Caribbean: A Junior Geography*. London, England: Macmillan Caribbean, 1985. 48 pp.

Comitas, L., and David Lowenthal, eds. *Slaves, Free Men, Citizens: West Indian Perspectives*. Garden City, N.Y.: Anchor, 1973. 340 pp.

Connolly, Quesnel A. "A Geographical Survey of Grand Cayman Island, British West Indies." Master's Thesis, Ohio State University, Columbus, Ohio, 1940.

Connor, G.A. "The Geographical Basis for Development of the West Indies." Master's Thesis, University of Western Ontario, London, Ontario, 1952.

Crist, Raymond E. "Static and Emerging Cultural Landscape on the Islands of St. Kitts and Nevis, B.W.I." *Economic Geography*, 25 (1949): 134-145.

Crist, Raymond E. "Cultural Dicotomy in the Island of Hispanola." *Economic Geography*, 28 (1952): 105-121.

Crompsey, Robert A. *Puerto Rico*. Harrisburg, Pa.: Stackpole Boos, 1973. 206 pp.

Crooker, Richard A. *Bermuda*. Broomwell, Pa.: Chelsea, 2002. 125 pp.

Cutteridge, J.O. *Geography of the West Indies and Adjacent Lands*. London, England: Thomas Nelson and Sons, 1947, 1962. 325 pp.

Dambaugh, Luella. "Jamaica: An Island in Transition." *Journal of Geography*, 52 (1953): 45-56.

Dambaugh, Luella. "Puerto Rico's Ambition." *Journal of Geography*, 50 (1961): 89-98.

Darkenwald, Gordon G. "Geographical Studies in Southwest Pinar del Rio." Ph.D. Dissertation, Clark University, Worcester, Mass., 1934.

Davis, William Morris. *The Lesser Antilles*. New York, N.Y.: American Geographical Society, 1926.

Davitaya, F.F. et al. "The Status of Geography in Cuba." *Soviet Geography*, 8, No. 1 (1967): 33-40.

De Booy, Theodore. "The Virgin Islands of the United States." *Geographical Review*, 4 (1917): 359-373.

De Booy, Theodore. "The Turks and Caicos Islands, British West Indies." *Geographical Review*, 6 (1918): 37-51.

De Haan, Leo J. "Small Islands in the Caribbean: The Last Remains of the Tropical Netherlands: The Netherlands Antilles and Aruba." *Tijdschrift voor Economische en Sociale Geographie*, 84, No. 5 (1993): 378-385.

Delobez, A. "La population et l'economie de la Martinique en 1960." *Information Geographique*, 1 (1964): 22-26.

Denevan, William M., ed. *Hispanic Lands and Peoples: Selected Writings of James J. Parsons*. Boulder, Colo.: Westview, Dellplain Latin American Studies, No. 23, 1989.

Desse, Michel. "Acteurs et dynamiques spatiales insulaires depuis la départementalisation en Guadeloupe, Martinique, Réunion." Ph.D. Dissertation, Université Paul Vabry, Montpellier, France, 1995.

Desse, Michel. "L'evolution contrastee du littoral de l'ile de la Martinique." *Mappemonde*, 55 (1999): 24-27.

Desse, Michel. *Les littoraux de la Caribe insulaire, des territoires en mutation. Habilitation a Diriger la Recherche*, Tome 2. Paris, France: University de Paris, 2004. 285 pp.

Desse, Michel. "La protection et la gestion des espaces littoraux en Guadeloupe et de la Martinique." *Etudes Caribeennes*, 1 (2004): 17-24.

Desse, Michel. "Haiti: Des littoraux en sursis." *Etudes Caribeennes*, 2 (2004): 17-31.

Desse, Michel. "Territorialisation et enclaves littorals dans les iles de la Caraibe." *Etudes Caribeennes*, 3 (2005): 9-14.

De Suze, J.A. *The New Trinidad and Tobago: A Descriptive Account of the Geography and History of Trinidad and Tobago*. London, England: Collins, 1965.

DeYoung, M. *Man and Land in the Haitian Economy*. Gainesville, Fla.: University of Florida Press, Latin American Monograph, No. 3, 1958.

Dodman, David, Duncan McGregor, and David Barker. "Beyond Caribbean Vulnerability: Towards Resistance and Resilience." Pp. 365-384 in *Global Change and Caribbean Vulnerability: Environment, Economy, and Society at Risk*, edited by Duncan McGregor, David Dodman, and David Barker. Mona, Jamaica: University of West Indies Press, 2009.

Doerr, Arthur H. "Environment and Economic Activities in Southwestern Puerto Rico." Pp. 185-212 in *Symposium on the Geography of Puerto Rico*, edited by Clarence Jones and Rafael Pico. Rio Piedras, Puerto Rico: University of Puerto Rico Press, 1955.

Doran, Edwin, Jr. "A Physical and Cultural Geography of the Cayman Islands." Ph.D. Dissertation, University of California, Berkeley, Calif., 1953.

Douglas, A.J.A. "The Cayman Islands." *Geographical Journal*, 95 (1940): 126-131.

Douglas, C.H. "Development Typologies, Strategies, and Priorities for Small Island States." *Caribbean Geography*, 13, No. 2 (2005): 110-113.

Dubesset, E. "Le tourism a Cuba: Une industrie en phine expansion." *Les Cahiers d'Outre Mer*, 48, No. 189 (1995): 35-54.

Dupuis, J. "Les paradoxes de Curacao. A travers les provinces de l'Empire Shell." *Les Cahiers d'Outre Mer*, 22 (1969): 63-74.

Dyer, Donald R. *Lesser Antilles*. Garden City, N.Y.: Doubleday, 1964.

Elbow, Gary S. "Regional Cooperation in the Caribbean: The Association of Caribbean States." *Journal of Geography*, 96, No. 1 (1997): 13-22.

Evans, F.C. *A First Geography of Jamaica*. London, England: Cambridge University Press, 1968.

Evans, F.C. *A First Geography of Trinidad and Tobago*. London, England: Cambridge University Press, 1968.

Evans, F.C. *The West Indies: A Geography of the West Indies*. London, England: Cambridge University Press, 1973. 126 pp.

Eyre, L. Alan. *A New Geography of the Caribbean*. London, England: G. Philip, 1962, 1964, 1968, and 1979.

Eyre, L. Alan. *Land and Population in the Sugar Belt of Jamaica*. Mona, Jamaica: University of West Indies, Department of Geography, Occasional Paper No. 1, 1966.

Eyre, L. Alan. "The Tropical National Parks of Latin America and the Caribbean: Present Problems and Future Potential." *Yearkbook, Conference of Latin Americanist Geographers*, 16 (1990): 15-33.

Fagg, John E. *Cuba, Haiti, and the Dominican Republic*. Englewood Cliffs, N.J.: Prentice-Hall, 1965.

Fairchild, F.R. "The Problem of Santo Domingo." *Geographical Review*, 10 (1920): 121-138.

Faugeres, M.I. "Observations sur le modele des versants dans la region des Pitens du Carbet (Martinique)." *Bulletin de la Association Geographes Francais*, 342, No. 3 (1966): 51-63.

Fleischmann, U. "Haiti: Kreolische Kultur und Politik." *Geographische Rundschau*, 49, No. 11 (1997): 651-655.

Floyd, Barry N. "Jamaica." *Focus*, 19, No. 2 (1968).

Floyd, Barry N. "The Two Faces of Jamaica." *Geographical Magazine*, 46 (1974): 424-431.

Floyd, Barry N. *Jamaica: An Island Microcosm*. New York, N.Y.: St. Martins Press, 1979. 164 pp.

Frankenhoff, Charles A., and L.T. Giulini. "Environment and Development: The Caribbean Region." *Ekistics*, 40 (1975): 281-286.

Gamble, Douglas W. "Re-Thinking the Region: South East Division, Association of American Geographers and the Caribbean." *Southeastern Geographer*, 44, No. 1 (2004): 8-14.

Giacottino, J.C. "La Barbade independante." *Les Cahiers d'Outre Mer*, 20 (1967): 209-227.

Giacottino, J.C. "Les petites Antilles britanniques." *Les Cahiers d'Outre Mer*, 23 (1970): 307-334.

Glassner, Martin I. "Jamaica: Land of Wood and Water and Names." *Places*, 1, No. 2 (1974): 7-13.

Gommersbach, Wilhelm. *Geschichte, Geographie und Bedeutung der Insel Trinidad.* Bonn, Germany: Bonner Kunstdruckerel A. Broch, 1907.

Gottmann, Jean. "The Isle of Guadeloupe." *Geographical Review*, 35 (1945): 182-203.

Graham, N.A., and K.L. Edwards. *Demographic, Economic, and Resource-Use Trends in 17 Caribbean Basin Countries*. Boulder, Colo.: Westview, 1984.

Greene, J. Maxwell. "Bermuda (Alias Somers Islands). Historical Sketch." *Bulletin of the American Geographical Society*, 33, No. 3 (1901): 220-242.

Hall, Robert B. "The Geography of the Republic of Haiti: A Study in the Structure and Evolution of Landscape." Ph.D. Dissertation, University of Michigan, Ann Arbor, Mich., 1927.

Hall, Robert B. "The Geography of the Republic of Haiti." *Scottish Geographical Magazine*, 46 (1930): 140-152.

Hanson, E.P. *Transformation: The Story of Modern Puerto Rico*. New York, N.Y.: Simon & Schuster, 1955.

Harris, David R. *Plants, Animals, and Man in the Outer Leeward Islands, West Indies*. Berkeley, Calif.: University of California Publications in Geography, No. 18, 1965.

Harris, W.N. "Puerto Rico: Island of Extremes." *Journal of Geography*, 48 (1949): 283-293.

Henshall, Janet D. "The Demographic Factor in the Structure of Agriculture in Barbados." *Transactions, Institute of British Geographers*, 38 (1966): 183-195.

Hough, Richard W.S. "In Search of Unity: Three Decades of Regional Integration in the Commonwealth Caribbean." Master's Thesis, Queen's University, Kingston, Ontario, 1992.

Hovey, Edmund O. "The Grand Soufriere of Guadeloupe." *Bulletin of the American Geographical Society*, 36, No. 9 (1004): 513-530.

Howard, Richard A., and J. Howard, eds. *Alexander Anderson's Geography and History of St. Vincent*. Cambridge, England: Linnean Society of London, 1983.

Hudson, Brian J. "The Changing Caribbean: Grenada's New International Airport." *Caribbean Geography*, 1, No. 1 (1983): 51-57.

Hudson, Brian J. "Putting Grenada on the Map." *Area*, 17, No. 3 (1985): 233-235.

Hudson, Brian J. "Geography in Colonial Schools: The Classroom Experience in West Indian Literature." *Geography*, 79, No. 4 (1994): 322-329.

Humlum, J. "St. Croix, St. Thomas, og St. Jan." *Kulturgeografi*, 16, No. 89 (1964): 53-80.

Inchaustegui, Cabral. *Geografia descriptive de la Republica Dominicana*. Ciudad Trujillo, Dominican Republic: n.p., 1957.

Indome, Kwesi. "Local Ecological Knowledge of Coral Reef Ecosystems: A Case Study of Bequia, St. Vincent and the Grenadines." Master's Thesis, Carleton University, Ottawa, Ontario, 2009.

Ishimo, Amani, Hugh Semple, and Elizabeth Thomas-Hope. "Population Mobility and the Survival of Small Farming in the Rio Grande Valley, Jamaica." *Geographical Journal*, 172, No. 4 (2006): 318-330.

Isnard, H. "La Reunion et la Martinique." *Les Cahiers d'Outre Mer*, 9 (1956): 58-69.

Jamaican Geographical Society. *Essays on Jamaica*. Kingston, Jamaica: University of the West Indies, 1970.

James, Preston E. "A Geographic Reconnaissance of Trinidad." *Economic Geography*, 3 (1927): 87-109.

James, Preston E. "Changes in the Geography of Trinidad." *Scottish Geographical Magazine*, 73 (1957): 157-166.

Jeffreys, T. *The West-India Atlas*. London, England: Robert Sayer and John Bennett, 1775.

Jenkins, Charles F. "Tortola, the Chief of the British Virgin Islands." *Bulletin of the Geographical Society of Philadelphia*, 21, No. 1 (1923): 1-21.

Jones, Clarence, and Rafael Pico, eds. *Symposium on the Geography of Puerto Rico*. Rio Piedras, Puerto Rico: University of Puerto Rico Press, 1955. 503 pp.

Joseph, P. "Le developpement durable dans les petites Antilles. Quelques reflexions generales." *Acta Geographica*, 179 (2007): 56-70.

Kapp, Kit. *The Printed Maps of Jamaica*. Kingston, Jamaica: Bolivar Press, 1968.

King, Ruby. "Geography in the Curriculum of Elementary Schools in Late Nineteenth Century Jamaica." *Caribbean Geography*, 5, No. 2 (1994): 136-143.

Kingsbury, Paul, and Thomas Klak. "Riddims of the Street, Beach, and Bureaucracy: Situating Geographical Research in Jamaica." *Southeastern Geographer*, 45, No. 2 (2005): 251-273.

Kingsbury, R.C. "The Virgin Islands." *Focus*, 11, No. 7 (1961).

Klak, Thomas, ed. *Globalization and Neoliberalism: The Caribbean Context*. Lanham, Md.: Rowman & Littlefield, 1997. 56 pp.

Kolivras, Korine, and Joseph L. Scarpaci. "Between Corporatism and Socialism: Navigating the Waters of International Education in the Dominican Republic and Cuba." *Journal of Geography*, 108, No. 3 (2009): 121-131.

Kucher, L. "Kubas Biolonbau-Revolution. Auskommen met den eigenen Ressourcen." *Geographische Rundschau*, 58, No. 12 (2006): 54-60.

Laguna Cruz, Jorge A. "Una aproximacion teorica al perfeccionamiento del proceso docente-educativo de la disciplina metodologia de la ensenanza de la geografia en el Instituto Superior Pedagogico de Holguin Cuba." *Revista Geografica*, 136 (2004) : 151-165.

Lambert, David, ed. *Atlas of Central America and the Caribbean*. New York, N.Y.: Macmillan, 1985. 144 pp.

Larsen, Soren. "Negril in the News: Content Analysis of a Contested Paradise." *Caribbean Geography*, 15, No. 1 (2008): 35-58.

Lasserre, Guy. "Marie-Galante." *Les Cahiers d'Outre Mer*, 3 (1950): 123-152.

Lasserre, Guy. "La Dominique et les derniers Caraibe insulaires." *Les Cahiers d'Outre Mer*, 6 (1952): 37-60.

Lasserre, Guy. "La Desirade." *Les Cahiers d'Outre Mer*, 10 (1957): 325-366.

Lasserre, Guy. *La Guadeloupe, Etude géographique*. 2 volumes. Bordeaux, France: Union francaise d'impression, 1961.

Leemann, A. "Trinidad." *Geographica Helvetica*, 24, No. 4 (1969): 191-194.

Leemann, A. "Barbados." *Geographica Helvetica*, 25 (1970): 130-135.

Lewis, Gordon. *The Growth of the Modern West Indies*. London, England: MacGibbon and Kee, 1968.

Lewis, Gordon. "An Introductory Note to the Study of the Virgin Islands." *Caribbean Studies*, 8, No. 2 (1968).

Liautaud, Andre. *Geographie d'Haiti.* Haiti: Editions Henri Deshcamps, 194?.

Long, George. *The Geography of America and the West Indies*. London, England: Society for the Diffusion of Useful Knowledge, 1841. 648 pp.

Lopez-Marrero, Tania, and N. Villanueva Colon. *Atlas Ambiental de Puerto Rico*. San Juan, Puerto Rico: University of Puerto Rico Press, 2006.

Lowenthal, David, ed. *The West Indian Federation: Perspectives on a New Nation*. New York, N.Y.: Columbia University Press, 1961. 142 pp.

Lowenthal, David. "Caribbean Views of Caribbean Land." *Canadian Geographer*, 5 (1961): 1-9.

Lowenthal, David, and Colin G. Clarke. "Island Orphans: Barbuda and the Rest." *Journal of Commonwealth and Comparative Politics*, 18, No. 3 (1980): 293-307.

Lyew-Ayee, Parris. "Redrawing the Boundaries of the Cockpit Country, Jamaica." *Caribbean Geography*, 14, No. 2 (2005): 102-115.

MacPherson, John. *Caribbean Lands*. Trinidad: Longman, 1981. 200 pp.

Magnon, A. "L'espace littoral a-tail toujours de la valeur? Reflexion a partir du cas de Saint-Martin (Petites Antilles)." *Norois*, 206 (2008): 37-52.

Marrero, Levi. *Geografia de Cuba*. Havana, Cuba: Editorial Selecta O'Reilly, 1951, 1957.

Massip, Salvador. *Introduccion a la geografia de Cuba*. Havana, Cuba: Fiallo y Hermanos, 1942.

McGregor, Duncan, and David Barker. "A Geographical Focus for Environment and Development in the Caribbean." Pp. 3-20 in *Environment and Development in the Caribbean: Geographical Perspectives*, edited by David Barker and Duncan McGregor. Kingston, Jamaica: University of West Indies Press, 1995.

McGregor, Duncan, David Dodman, and David Barker, eds. *Global Change and Caribbean Vulnerability*. Mona, Jamaica: University of West Indies Press, 2009.

Menanteau, L., and J.R. Vanney. *Atlas cotier du nord-est d'Haiti*. Port-au-Prince, Haiti: Ministere de la Culture, 1997.

Merrill, Gordon C. "The Survival of the Past in the West Indies." Pp. 17-34 in *The West Indies Federation*, edited by David Lowenthal. Washington, D.C.: American Geographical Society Research Series, No. 23, 1961.

Mitchell, Harold P. *Caribbean Patterns*. New York, N.Y.: Wiley, 1972. 583 pp.

Momsen, R.P., ed. *Geographical Analysis for Development in Latin America and the Caribbean*. Chapel Hill, N.C.: University of North Carolina, Department of Geography, Conference of Latin Americanist Geographers, Publications, 1975.

Morrissey, Michael, and David Barker. *A Sense of Place I: Introducing Caribbean Geography*. London, England: Oxford University Press, 1991. 128 pp.

Murray, Helen C. "Border Crossings: Africana, the Caribbean, Pan-Africanism." Master's Thesis, Carleton University, Ottawa, Ontario, 1998.

Nash, James. "From Periphery to the Core: EPZ's in the Caribbean." *Bulletin, Association of North Dakota Geographers*, 45 (1995): 65-78.

National Atlas of Jamaica. Kingston, Jamaica: Town Planning Department, Ministry of Finance and Planning, 1971.

Netzer, Donald L. "Climate and Agricultural Cropland Use in Puerto Rico." Ph.D. Dissertation, University of Illinois, Urbana, Ill., 1953.

Nicolas, Thierry. "Les Antilles francais entre insularite et continentalite." *Geographie et Cultures*, 36 (2000): 39-56.

Niddrie, David L. *Land Use and Population in Tobago: An Environmental Study*. London, England: Geographical Publications, 1961.

Niddrie, David L. "The Caribbean Islands Today." *Journal of the Institute of Bankers* (June 1963): 1-11.

Niddrie, David L. *Tobago*. New York, N.Y.: Methuen, 1981. 243 pp.

Nietschmann, Bernard. *The Caribbean Edge: The Coming of Modern Times to Isolated People and Wildlife*. New York, N.Y.: Bobbs-Merrill, 1979. 280 pp.

Nordlohne, E. "The Netherlands Antilles." *Tijdschrift voor Economische en Sociale Geographie*, 47 (1956): 167-170.

Norton, Ann, ed. *Essays on Jamaica*. Mona, Jamaica: University of West Indies, Department of Geography, 1970.

Nuevo atlas nacional de Cuba. Havana, Cuba: Instituto de Geografia de la Academia de Ciencias de Cuba, 1989.

Nunez Jimenez, Antonio. *Geografia de Cuba*. Havana, Cuba: Instituto Cubano del Libro, 1959 and 1972.

Olschki, Leonardo. "The Colombian Nomenclature of the Lesser Antilles." *Geographical Review*, 33 (1943): 397-414.

Pagney, Francoise et al. "Espace a risques en milieu tropical insulaire." *Acta Geographica*, 125 (2001): 7-26.

Papy, L. "Paysages et gens de la Martinique." *Les Cahiers d'Outre Mer*, 2 (1949): 333-370.

Pearcy, George E. *The West Indian Scene*. Princeton, N.J.: van Nostrand, 1965. 136 pp.

Pereira, Paul. *Geographie d'Haiti*. Port-au-Prince, Haiti: n.p., 1938.

Pichardo y Tapia, Esteban. *Geografia de la isla de Cuba*. Havana, Cuba: D.M. Soler, 1854-1855.

Pico, Rafael. "Geographic Regions of Puerto Rico." Ph.D. Dissertation, Clark University, Worchester, Mass., 1938.

Pico, Rafael. "Comparisons and Contrasts in the Greater Antilles." *Hispanic American Studies*, 2 (1941): 132-142.

Pico, Rafael. *The Geographic Regions of Puerto Rico*. Rio Piedras, Puerto Rico: University of Puerto Rico Press, 1950. 256 pp.

Pico, Rafael. "The Commonwealth of Puerto Rico." *Focus*, 14, No. 2 (1963).

Pico, Rafael. *Nueva Geografia de Puerto Rico: Fisica, Economia, y Social*. San Juan, Puerto Rico: Editorial Universitaria, 1969.

Pico, Rafael. *The Geography of Puerto Rico*. Chicago, Ill.: Aldine, 1974.

Poey, Felipe. *Geografia fisica y politica de la isla de Cuba*. Havana, Cuba: Imprenta y papeleria de la viude de Barcina, 1860.

Potter, Amy. "Voodoos, Zombies, and Mermaids: U.S. Newspaper Coverage of Haiti." *Geographical Review*, 99, No. 2 (2009): 208-230.

Potter, Robert B. *Barbados*. Oxford, England: Clio, 1987. 356 pp.

Potter, Robert B. "Basic Needs and Development in the Small Island States of the Eastern Caribbean." Pp. 92-116 in *The Development Process in Small Island States*, edited by Douglas G. Lockhart, David Drakakis-Smith, and John Schembri. London, England: Routledge, 1993.

Preiswerk, Roy. *Regionalism and the Commonwealth Caribbean*. Port of Spain, Trinidad: Institute of International Relations, University of the West Indies, 1969.

Price, Edward T. "Notes on the Geography of Barbados." *Journal of the Barbados Museum and Historical Society*, 29 (1962): 119-154.

Pruna Santa Cruz, Manuel. *Geografia elemental de la isla de Cuba*. Havana, Cuba: J.L. Rodriguez, 1913.

Puerto Rico. Garden City, N.Y.: American Geographical Society, 1968. 64 pp.

Questel, Adrien. *Geographie generale de la Guadeloupe et dependances*. Paris, France: Le Charles, 1941. 327 pp.

Ratter, B., and W. Sahr, eds. *Land, Sea, and Human Effort in the Caribbean*. Hamburg, Germany: Beitrage zur Geographischen Regionalforschung in Lateinamerika, Volume 10, 1997.

Redon, M. "Saint-Martin/Sint Maarten, une petite ile devisee pour de grande enjeus." *Les Cahiers d'Outre Mer*, 59 (2006): 233-266.

Revert, E. "Perspectives caraibes." *Les Cahiers d'Outre Mer*, 1 (1948): 334-347.

Revert, E. *La Martinique. Etude geographique et humaine*. Paris, France: Nouvelles ed.s Latines, 1949.

Revert, E. "Problemes de Geographie antillaise." *Les Cahiers d'Outre Mer*, 3 (1950): 1-27.

Revert, E. "La Republique d'Haiti." *Revue de Geographie de Lyon*, 27 (1952): 309-316.

Revert, E. *Les Antilles*. Paris, France: A. Colin, 1954. 220 pp.

Revert, E. *Entre les deux Ameriques: Le monde Caraibe*. Paris, France: Les ed.s Francaises, 1958.

Richardson, Bonham C. "Catastrophes and Change in St. Vincent." *National Geographic Research*, 5 (1989): 111-125.

Robequain, C. "Saint-Bartheleny, terre francaise." *Les Cahiers d'Outre Mer*, 2 (1949): 14-37.

Sagawe, Thorsten. "Deforestation and the Behavior of Households in the Dominican Republic." *Geography*, 76 (1991): 304-314.

Sanguin, A.L. "Saint-Martin, les mutations d'une ile franco-neerlandaise des Antilles." *Les Cahiers d'Outre Mer*, 35 (1972): 123-140.

Santana, Deborah B. "Puerto Rico's Operation Bootstrap: Colonial Roots of a Persistent Model for 'Third World' Development." *Revista Geografica*, 124 (1998): 87-116.

Scarpaci, Joseph L., and A. Portela. *Cuban Landscapes*. New York, N.Y.: Guilford, 2009.

Sealey, Neil E. *Bahamian Landscapes: An Introduction to the Geography of the Bahamas*. London, England: Collins Caribbean, 1985.

Sealey, Neil E. *Caribbean World: A Complete Geography*. Cambridge, England: Cambridge University Press, 1992. 256 pp.

Segre, R. "Die strukturellen Veranderungen in Kuba." *Urbs et Region*, 31 (1984): 189-216.

Sellers, J.R., and P. Molen van Ee. *Maps and Charts of North America and the West Indies, 1750-1789*. Washington, D.C.: Library of Congress, 1981. 495 pp.

Senior, Michael. *Tropical Lands: A Human Geography*. London, England: Longman, 1979. 495 pp.

Shaw, Earl B. "Geographic Studies in the Virgin Islands of the United States." Ph.D. Dissertation, Clark University, Worchester, Mass., 1933.

Shaw, Earl B. "St. Thomas: The Keystone of the Antilles." *Journal of Geography*, 32 (1934): 131-139.

Skelton, Tracey. *Introduction to the Pan-Caribbean*. New York, N.Y.: Hodder-Arnold, 2004. 200 pp.

Smith, Roger C. "The Maritime Geography of the Cayman Islands." *Caribbean Geography*, 1, No. 4 (1984): 247-255.

Sperling, Jonathan. "Understanding Census Geography in the Tiger Data Base: The Case of Puerto Rico." *Middle States Geographer*, 24 (1991): 121-126.

Speth, W.W. *Notes on the Coastal Geography of Trinidad.* Berkeley, Calif.: University of California, Department of Geography, 1962.

Street, John M. "Historical and Economic Geography of the Southwest Peninsula of Haiti." Ph.D. Dissertation, University of California, Berkeley, Calif., 1959.

Stuart, R. "Haiti or Hispanoola?" *Journal of the Royal Geographical Society*, 48 (1878): 234-274.

Symanski, Richard, and Nancy Burley. "The Jewish Colony of Sosua." *Annals of the Association of American Geographers*, 63 (1973): 366-378.

Taillefer, F. "Les Bermudes, esquisse geographique." *Les Cahiers d'Outre Mer*, 11 (1958): 101-116.

Tata, Robert J. *Haiti: Land of Poverty.* Washington, D.C.: University Press of America, 1982. 159 pp.

Thomas, L. "La Dominique et les derniers Caraibes insulaires." *Les Cahiers d'Outre Mer*, 6 (1953): 36-60.

Thomas-Hope, Elizabeth, ed. *Perspectives on Caribbean Regional Identity.* Liverpool, England: University of Liverpool, Center for Latin American Studies, No. 11, 1984. 134 pp.

Toro Sugranes, Jose A. *Nuevo atlas de Puerto Rico.* Rio Piedras, Puerto Rico: Editorial Edil, 1982.

Ventura y Fernandez, Victorio R. *Geografia ilustrada de al isla de Cuba.* Havana, Cuba: n.p., 1897.

Verin, P. "Sainte-Lucie et ses derniers Caraibes." *Les Cahiers d'Outre Mer*, 12 (1959): 341-361.

Verin, P. "Les Caraibes a Sainte Lucie depus les contacts coloniaux." *Nieuwe West-Indische Gids*, 44 (1961): 66-82.

Versluis, Anna J. "We All Live Downstream: Disaster, Land Change, and Reciprocity in the Haitian Watershed." Ph.D. Dissertation, Clark University, Worcester, Mass., 2008.

Verteuil, L.A.A. *Trinidad: Its Geography, Natural Resources, Administration, Present Condition, and Prospects.* London, England: Cassell, 1858; second edition, 1884.

Von Royen, William. "A Geographical Reconnaissance of the Cibao of Santo Domingo." *Geographical Review*, 28 (1938): 556-572.

Watson, J. Wreford, J. Oliver, and C.H. Foggo. *Bermuda: A Geography*. London, England: Collins, 1964.

Watts, David J. *The West Indies: Patterns of Development, Culture, and Environmental Change Since 1492*. Cambridge, England: Cambridge University Press, 1987. 609 pp.

Wehrhahn, Rainer, and Sonke Widderich. "Tourismas als Entwicklungs-faktor im Kubanishcen Transformationsprozess." *Erdkunde*, 54, No. 1 (2000): 93-107.

West, Robert C., and John P. Augelli. *Middle America: Its Lands and Peoples*. Englewood Cliffs, N.J.: Prentice-Hall, 1966, 1976, and 1989.

Weyl, R. "Bau und Bild der Kleinen Antillen." *Geographische Rundschau*, 15, No. 3 (1963): 103-107.

Whitbeck, Florence. "A Geographical Interpretation of Porto Rico." Ph.D. Dissertation, University of Wisconsin, Madison, Wis., 1927.

Whitbeck, R.H. "The Lesser Antilles: Past and Present." *Annals of the Association of American Geographers*, 23 (1933): 21-26.

Whitehead, Neil L. "Indigenous Cartography in Lowland South America and the Caribbean." Pp. 301-326 in *The History of Cartography*, edited by David Woodward and M. Lewis. Chicago, Ill.: University of Chicago Press, 1998.

Wiley, James. "Dominca, Grenada, and the Non-Traditional Agricultural Exports (NTAE) Imperative." *Yearbook, Conference of Latin Americanist Geographers*, 25 (1999): 77-88.

Wilhelmy, H. "Dreissig Jahre Kubanisches 'Entwicklungmodels 1. Beobachtungen auf zwei Reisen 1959 und 1991." *Die Erde*, 123 (1992): 321-328.

Wilken, Gene C. "Future Caribbean Donor Landscapes: A Geographic Interpretation of Contemporary Trends." *Caribbean Geography*, 3, No. 4 (1992): 215-222.

Winsemius, J. "Some Notes on St. Maartens, Bonaire, and Saba (Netherlands Antilles)." *Tijdschrift voor Economische en Sociale Geographie*, 54 (1963): 202-208.

Yearman, Keith. "Navassa Island: A Look at Hell." *Bulletin of the Illinois Geographical Society*, 40, No. 1 (1998): 2-7.

Zapata-Rivera, Lucila. "Geography of Puerto Rico: A Text for Secondary Schools." Ph.D. Dissertation, Columbia Teachers College, New York, N.Y., 1958.

Part 2

Cultural and Social Geography

Aagensen, David, and Jennifer Palucle. "International Tourism Trends in Cuba." *Geographical Bulletin*, 43, No. 1 (2001): 47-58.

Abreau, Leonarda. "Impacto del Turismo en el Litoral de Dominicana." *Revista Geografica*, 126 (1999): 167-182.

Adams, John E. "From Landsmen to Seamen: The Making of a West Indian Community." *Revista Geografica*, 88 (1978): 151-166.

Ahsan, Syed R. "East Indian Agricultural Settlements in Trinidad: A Study in Cultural Geography." Ph.D. Dissertation, University of Florida, Gainesville, Fla., 1963.

Ainsley, W. Frank. "Evolution of the Chattel House: Folk Housing of Barbados." *Transactions, Pioneer America Society*, 19 (1996): 31-40.

Akman, Gerri. "Both Sides Now: Gender Relations in Credit and Agricultural Cooperatives in Rural Haiti." Ph.D. Dissertation, McGill University, Montreal, Quebec, 1993.

Albers, B. "Planning in Paradise: Developing an Environmentally South Resort in the Caribbean." *GeoInfosystems*, 1, No. 10 (1991): 26-38.

Albuquerque, Klaus de, and Jerome L. McElroy. "West Indian Migration to the United States Virgin Islands: Demographic and Socio-Economic Consequences." *International Migration Review*, 16 (1982): 61-101.

Albuquerque, Klaus de, and Jerome L. McElroy. "Alternative Tourism and Sustainability." Pp. 23-32 in *Island Tourism*, edited by M.V. Conlin and T. Baum. Chichester, England: John Wiley, 1990.

Albuquerque, Klaus de, and Jerome L. McElroy. "Caribbean Small State Tourism Styles and Sustainable Strategies." *Environmental Management*, 16 (1992): 619-632

Albuquerque, Klaus de, and Jerome L. McElroy. "Tourism Development in Small Islands: St. Maartens/St. Martin and Bermuda." Pp. 70-89 in *Environment and Development in the Caribbean: Geographical Perspectives*, edited by David Barker and Duncan McGregor. Kingston, Jamaica: University of the West Indies Press, 1995.

Albuquerque, Klaus de, and Jerome L. McElroy. "Tourism and Crime in the Caribbean." *Annals of Tourism Research*, 26, No. 4 (1999): 968-984.

Andre, B. "Demographic Transition in Haiti (in French)." *Bulletin de l'Association de Geographes Francais*, 1 (2011): 83-104.

Arbona, Sonia, and Markku Loytenen. "AIDS Epidemic in Puerto Rico: Ideas for Geography Teaching." *Fennia*, 175, No. 2 (1997): 125-138.

Archer, Ewart D.G. "Effects of the Tourist Industry in Barbados, West Indies." Ph.D. Dissertation, University of Texas, Austin, Tex., 1980.

Atherly, Kenneth. "The Application of Tourist-Based Research to Coastal Management in Barbados." Master's Thesis, McGill University, Montreal, Quebec, 1986.

Atkin, Robert A. "Surgiderp de Batabana: A Cuban Fishing Village." Master's Thesis, University of Tennessee, Knoxville, Tenn., 1952.

Audebert, Cedric, et Michel Desse. "L'articulation entre migrations, mobilites et constructions identitaires: Une nouvelle approche de l'insertion de la Caribe dans Les reseaux de la mondialisation." *Etudes Caribeennes*, 8 (2007).

Augelli, John P. "San Lorenzo: A Case Study of Recent Migration in Interior Puerto Rico." *American Journal of Economics and Sociology*, 11, No. 2 (1951): 155-160.

Augelli, John P. "Rural Settlement Types of Interior Puerto Rico: Sample Studies from the Upper Coiza Basin." *Journal of Geography,* 51 (1952): 1-12.

Augelli, John P. "Rural Settlement Types of Interior Puerto Rico: Sample Studies from the Upper Coiza Basin." Pp. 325-336 in *Symposium on the Geography of Puerto Rico*, edited by Clarence Jones and Rafael Pico. Rio Piedras, Puerto Rico: University of Puerto Rico Press, 1955.

Augelli, John P. "The Mainland-Rimland Concept of Culture Regions in Middle America." *Annals of the Association of American Geographers*, 52, No. 2 (1962): 119-129.

Augelli, John P., and Harry Taylor. "Race and Population Patterns in Trinidad." *Annals of the Association of American Geographers*, 50, No. 2 (1960): 123-138.

Austin, Susan. "A Place of Globalization, Resistence, and Identity: A Case Study of the Santa Rosa Carib Community in Trinidad." Master's Thesis, Syracuse University, Syracuse, N.Y., 2004.

Azar, Derek, and David Rain. "Identifying Population Vulnerable to Hydrological Hazards in San Juan, Puerto Rico." *GeoJournal*, 69, No. 1-2 (2007): 23-43.

Bailey, Adrian J., and Mark Ellis. "Going Home: The Migration of Puerto Rican-Born Women from the United States to Puerto Rico." *Professional Geographer*, 45, No. 2 (1993): 148-158.

Bailey, Wilma, and D.R. Phillips. "Spatial Patterns of Use of Health Services in the Kingston Metropolitan Area, Jamaica." *Social Science and Medicine*, 30, No. 1 (1989): 1-12.

Bailey, Wilma, and Dorian Powell. "Patterns of Contraceptive Use in Kingston and St. Andrew, Jamaica, 1970-1977." *Social Science and Medicine*, 16, No. 19 (1982): 1675-1683.

Bailey, Wilma, H. Wynter, and A. Lee. "Race Integration and Family Planning in Trinidad and Tobago." *GeoJournal*, 26, No. 1 (1992): 13-20.

Baldwin, Jeff. "Tourism Development, Wetland Degradation, and Beach Erosion in Antigua, West Indies." *Tourism Geographies*, 2, No. 2 (2000): 193-218.

Barker, David. "A Periphery in Genesis and Exodus: Reflections on Rural-Urban Relations in Jamaica." Pp. 294-322 in *The Geography of Urban-Rural Interactions in Development Countries*, edited by Robert B. Potter and Timothy Unwin. London, England: Routledge, 1989.

Barrett, Elinore M. "Colonization of the Santo Domingo Valley." *Annals of the Association of American Geographers*, 64, No. 1 (1974): 34-53.

Barsotti, Natasha. "Social Processes and Environmental Factors Contributing to the Development of the Steelband in Port of Spain, Trinidad." Master's Thesis, University of Texas, Austin, Tex., 1996.

Baxter, A.E. "The Diffusion of Innovations: Soil Conservation Techniques, the Yallah Valley, Jamaica." *Jamaica Journal*, 9 (1975): 41-56.

Benitez, Javiar. "Spatial Diffusion of Multimedia Technologies Among Educators: A Case Study of Puerto Rico." Master's Thesis, Binghamton University, Binghamton, N.Y., 2009.

Benoist, J. "Les relations d'une societe et d'un milieu geographique. La plantation antillaise." Pp. 493-500 in *Etudies de geographie tropicale offertes a Pierre Gourou*. Paris, France: Mounton, 1972.

Bergman, John F., and J. Gordon Nelson. "Jamaica: Prospects for a New Nation." *Journal of Geography*, 63, No. 3 (1964): 116-122.

Berleant-Schiller, Riva. "Hidden Places and Creole Forms: Naming the Barbudan Landscape." *Professional Geographer*, 43, No. 1 (1991): 92-101.

Berry, M. Victoria. "Origin of Medicinal Plants and Knowledge of Their Use: Montserrat, West Indies." Ph.D. Dissertation, University of Tennessee, Knoxville, Tenn., 1999.

Berry, M. Victoria. "Geographic Origins of Medicinal Plants in Monserrat, West Indies, Pre-1632 to Present." *Southeastern Geographer*, 33, No. 1 (2003); 75-88.

Berry, M. Victoria. "Exploring the Potential Contributions of Amerindians to West Indian Folk Medicine." *Southeastern Geographer*, 45, No. 2 (2005): 239-250.

Bischoff, Carola. "Tourism in Dominica: Influence on the Retail and Service Sector of Roseau." *Caribbean Geography*, 10, No. 1 (1999): 19-36.

Bischoff, Carola. *Kruezfahrt und Okotourismus in Dominica. Auswirkungen auf die Nutzungsstruckture der Haupstadteines Karibischen Mikrostaates*. Munster, Germany: Munstersche Geographische Arbeiten, Volume 44, 2000.

Blaunt, James M. et al. "A Study of the Cultural Determinants of Soil Erosion and Conservation in the Blue Mountains of Jamaica." *Social and Economic Studies*, 8 (1959): 403-420.

Blue, Sarah A. "State Policy, Economic Crisis, Gender, and Family Times: Determinants of Family Remittances to Cuba." *Economic Geography*, 80, No. 1 (2004): 63-82.

Blue, Sarah A. "Socio-Economic Costs of Globalization: Remittances, the Informal Economy, and Rising Inequality in Contemporary Cuba." Ph.D. Dissertation, University of California, Davis, Calif., 2004.

Blue, Sarah A. "Cuban Medical Internationalism: Domestic and International Impacts." *Journal of Latin American Geography*, 9, No. 1 (2010): 31-49.

Blume, Helmut. "Die britischer Inseln uber dem Winde (Kleine, Antillen). Grundbesitz und Betriebsformen in ihrem Einfluss auf das Bild der Kulturlandschaft." *Erdkunde*, 15, No. 4 (1961): 265-286.

Blume, Helmut. "Kulturgeographische Divergenzen im Landschaftsbilde Hispaniolas." *Acta Geographica*, 20 (1968): 31-45.

Blume, Helmut. "Die Kulturraumliche Gliederung der Antillen." Pp. 335-348 in *Wirtschafts- und Kulturraume der Aussereuropaischen Welt. Festschrift fur Albert Kolb*, edited by Gunter Borchert et al. Hamburg, Germany: Hamburger Geographische Studien, Volume 24, 1971.

Boswell, Thomas D. "Municipio Characteristics as Factors Affecting Individual Migration in Puerto Rico: 1935-1940 and 1955-1960." Ph.D. Dissertation, Columbia University, New York, N.Y., 1973.

Boswell, Thomas D. "Net Migration on a Munipio Scale in Puerto Rico, 1960-1970." *Proceedings, Conference of Latin Americanist Geographers* (1974): 75-82.

Boswell, Thomas D. "Inferences Concerning Intermunicipio Migrations in Puerto Rico: 1955-1960." *Journal of Tropical Geography*, 45 (1977): 1-11.

Boswell, Thomas D. "Distance and Migration Selectivity in Puerto Rico Prior to Economic Development." *Revista Geografica*, 85 (1977): 189-203.

Boswell, Thomas D. "An Inventory of Migration Research Dealing with Caribbean Topics." *Proceedings, Conference of Latin Americanist Geographers* (1981): 100-118.

Boswell, Thomas D. "In the Eye of the Storm: The Context of Haitian Migration to Miami, Florida." *Southeastern Geographer*, 23, No. 2 (1983): 57-77.

Boswell, Thomas D. "The Migration and Distribution of Cubans and Puerto Ricans Living in the United States." *Journal of Geography*, 83, No. 2 (1984): 65-72.

Boswell, Thomas D., and A.K. Ohibwa. "A Comparison of Net Migration Patterns and Their Spatial Correlates in Puerto Rico and the Bahamas during the 1960s." *Geographical Survey*, 8 (1979): 16-29.

Bothwell, Lyman D. "The Geographic Factor in the Distribution, Trends, and Peaks of the Population of Puerto Rico, 1900-1930." *Papers, Michigan Academy of Science, Arts, and Letters*, 22 (1936): 321-327.

Bounds, John H. "The Bahamas Tourism Industry: Past, Present, and Future." *Revista Geografica*, 88 (1978): 167-220.

Bowden, M.J. *Hurricane in Paradise: Perception and Reality of the Hurricane Hazard in the Virgin Islands*. St. Thomas, U.S. Virgin Islands: Island Resources Foundation, 1974.

Britton, Edward C. "Peasant Subsistence in Northwestern Haiti: Geography, Cultural Ecology and Rural Development." Ph.D. Dissertation, Louisiana State University, Baton Rouge, La., 1993.

Britton, Robert A. "International Tourism and Indigenous Development Objectives: A Study with Spatial References to the West Indies." Ph.D. Dissertation, University of Minnesota, Minneapolis, Minn., 1978.

Brookfield, Harold. "The Environment of Informality: A Comparison Between Caribbean and Pacific." Pp. 257-274 in *Food, Shelter, and Transport in Southeast Asia and the Pacific*, edited by Peter J. Rimmer, David Drakakis-Smith, and Terrence G. McGee. Canberra, Australia: Australian National University, Research School of Pacific Studies, Department of Geography, Publication HG/12, 1978.

Bunce, Michael. "The Leisuring of Rural Landscapes in Barbados: New Spatialities and the Implications for Sustainability in Small Island States." *Geoforum*, 39, No. 2 (2008): 969-979.

Burne, Maurice. "Tourism and Environment in Guadeloupe and Martinique." Pp. 63-74 in *Sustainable Tourism in Islands and Small States*, edited by Lina Briguglio et al. London, England: Pinter, 1996.

Butler, Richard W. "Tourism Development in Small Islands: Past Influences and Future Directions." Pp. 71-91 in *The Development Process in Small Island States*, edited by Douglas G. Lockhart, David Drakakis-Smith, and John Schembri. London, England: Routledge, 1993.

Byrne, A. Roger. "Man and the Variable Vulnerability of Island Life: A Study of Recent Vegetation Change in the Bahamas." Ph.D. Dissertation, University of Wisconsin, Madison, Wis., 1972.

Byron, Margaret. "The Caribbean-Born Population in 1990s Britain: Who Will Return?" *Journal of Ethnic and Migration Studies*, 25, No. 2 (1999): 285-301.

Byron, Margaret. "Return Migration to the Eastern Caribbean: Comparative Experiences and Policy Implications." *Social and Economic Studies*, 49, No. 4 (2000): 155-188.

Byron, Margaret, and Stephanie London. "A Comparative Study of Caribbean Return Migration from Britain and France: Towards a Context-Dependent Explanation." *Transactions, Institute of British Geographers*, 21, No. 1 (1996): 91-104.

Camarena, Andres A. "La Estrategia de los Asentamientos Humanos en la Republica Dominicana: La Planificacion regional como instrumento estralegico en la distribution de la poblacion en el territorio." *Revista Geografica*, 91-92 (1980): 41-68.

Campbell, Douglas A. "St. Thomas: Sequent Occupance and the Development of Tourism." Master's Thesis, University of Miami, Miami, Fla., 1977.

Campbell, Rosemary. "Adaptation in Adversity: Peasant Response to Environmental and Economic Constraints in Nevis, West Indies." Master's Thesis, McGill University, Montreal, Quebec, 1974.

Canally, Culum. "The Social Construction of Cuba's Tourist Destination Image." Ph.D. Dissertation, University of Waterloo-Wilfrid Laurier University Graduate Program, Waterloo, Ontario, 2009.

Cant, A.G. "A Case Study in Outmigration: Movement of Adult Farm Workers from the Selwyn-Ashburton Electoral District During the Period 1963-1966." Pp. 41-48 in *Proceedings, Fifth New Zealand*

Geography Conference, edited by A.G. Anderson. Auckland, New Zealand: New Zealand Geographical Society, 1967.

Carlson, F.A. "American Settlement in the Isla de Pinos, Cuba." *Geographical Review*, 32, No. 1 (1942): 21-35.

Carpenter, Bruce R. "Puerto Rico's Planned Development of Tourism." Ph.D. Dissertation, American University, Washington, D.C., 1964.

Cazes, G. "Le developpement du tourisme a la Martinique." *Les Cahiers d'Outre Mer*, 21 (1968): 225-256.

Cazes, G. "Problemes de population et perspectives economiques en Martinique et en Guadeloupe." *Les Cahiers d'Outre Mer*, 23 (1970): 379-424.

Chardon, Jean-Pierre. *Geographie antillaise: La population d'une petite Antille: La Dominique*. Pointe-a-Pitre, F.W.I.: Groupe universitaire de recherches inter-Caraibes, Etudes et documents, No. 11, 1971.

Chaves, Antonio F. "A Dot Map of the Distribution of Population in Cuba." *Acta Americana*, 5 (1947): 261-266.

Chaves, Antonio F. "Aspectos geograficos del poblaciamento de Puerto Rico." *Revista Geografica*, 5-8 (1949): 57-68.

Chaves, Antonio F. *La distribution de la poblacion en Puerto Rico*. San Juan, Puerto Rico: Universidad de Puerto Rico, 1949.

Chivallon, Christine. "Tradition et modernite dans le monde paysan martiniques: Approche ethno-geographique." Ph.D. Dissertation, Universite de Bordeaux, Bordeaux, France, 1992.

Chivallon, Christine. "Diversite des fonctionnements des maisonnees rurales des mornes Martiniquais: Essai de classification." *Les Cahiers d'Outre Mer*, 45 (1993): 105-114.

Chivallon, Christine. "Les paysans martiniquais." *Geographie et Cultures*, 7 (1993): 3-26.

Chivallon, Christine. *Espace et identite a la Martinique. Paysannerie des mornes et reconquete colletive 1840-1960*. Paris, France: CNRS-Editions, 2002.

Chivallon, Christine. *La Diaspora noire des Ameriques. Experiences et theories a partir de la Caraibe*. Paris, France: CNRS ed.s, 2004. 258 pp.

Chivallon, Christine. "Space, Memory, and Identity in Martinique: The Beautiful Story of Providence (in French)." *Annales de Geographie*, 113, No. 4-5 (2004): 400-424.

Clarke, Colin G. "Land Tenure and the Family in Four Selected Communities in Jamaica." *Social and Economic Studies*, 1, No. 4 (1953): 81-118.

Clarke, Colin G. "Caste Among Hindus in a Town in Trinidad: San Fernando." Pp. 165-199 in *Caste in Overseas Indian Communities*, edited by B.M. Schwartz. San Francisco, Calif.: Chandler, 1967.

Clarke, Colin G. "Population Problems in the Caribbean." *Revista Geografica,* 75 (1971): 165-182.

Clarke, Colin G. "The Political Ecology of a Tow in Trinidad." Pp. 788-801 in *International Geography 1972*, edited by W.P. Adams and F.M. Helleiner. Montreal, Quebec: International Geographical Congress, 1972.

Clarke, Colin G. "Residential Segregation and Intermarriage in San Fernando, Trinidad." *Geographical Review*, 61, No. 2 (1972): 198-218.

Clarke, Colin G. "Pluralism and Stratification in San Fernando, Trinidad." Pp. 53-70 in *Social Patterns in Cities*, edited by B.C. Clark and M.B. Gleave. London, England: Institute of British Geographers, Special Publication No. 5, 1973.

Clarke, Colin G. "Jamaica Overflows." *Geographical Magazine*, 46 (1974): 615-619.

Clarke, Colin G. "Scale Factors and Ethnic Patterns in Trinidad." *Geographical Society Journal* (University of Liverpool), 4 (1976): 30-34.

Clarke, Colin G. "Pluralism and Plural Societies: Caribbean Perspectives." Pp. 51-86 in *Geography and Ethnic Pluralism*, edited by Colin G. Clarke, David Ley, and Ceri Peach. London, England: George Allen and Unwin, 1984.

Collymore, J.M. "The Impact of Natural Hazards on the Health Sector in the Caribbean: Implications for Health Sector Planning." *GeoJournal*, 26, No. 1 (1992): 61-64.

Condon, S.A., and P.E. Ogden. "Emigration from the French Caribbean: The Origins of an Organized Migration." *International Journal of Urban and Regional Research*, 15, No. 4 (1991): 505-523.

Connelly, John. "Anguilla: The Tourism Trajectory in an Island Micro-state." *Caribbean Geography*, 4, No. 2 (1993): 131-137.

Conway, Dennis. "Caribbean International Mobility Transitions." *Boletin Latino-Americanos y del Caribe*, 46 (1989): 17-47.

Conway, Dennis. "Migration in the Caribbean." *Yearbook, Conference of Latin Americanist Geographers*, 17-18 (1990): 91-98.

Conway, Dennis. "Why Barbados Has Exported People: International Mobility as a Fundamental Force in the Creation of Small Island Society." Pp. 274-308 in *Ethnicity, Race, and Nationality in the Caribbean*, edited by J.M. Carrion. San Juan, Puerto Rico: University of Puerto Rico, Institute of Caribbean Studies, 1997.

Conway, Dennis. "Caribbean Transnational Migration Behaviour: Reconceptualizing Its Strategic Flexibility." *Population, Space, and Place*, 13, No. 6 (2007): 415-431.

Conway, Dennis, Adrain J. Bailey, and Mark Ellis. *The International Circulation of Puerto Rican Women: Is It Structurally Determined?* San German, Puerto Rico: Inter American University of Puerto Rico, Caribbean Institute and Study Center for Latin America, Working Paper 47, 1991.

Conway, Dennis, and J. Brown. "The Changing Relations of Intraurban Relocation and Structure: Towards an Evolutionary Model of Low-Income Migrant Relocation in Latin America and the Caribbean." *Latin American Research Review*, 15, No. 3 (1980): 95-126.

Conway, Dennis, and Corrine Glesne. "Rural Livelihood, Return Migration, and Remittances in St. Vincent." *Yearbook, Conference of Latin Americanist Geographers*, 12 (1986): 3-11.

Conway, Dennis, and Robert B. Potter. "Caribbean Transnational Return Migrants as Agents of Change." *Geography Compass*, 1, No. 1 (2007): 25-45.

Conway, Dennis, Robert B. Potter, and Godfrey St. Bernard. "Repetitive Visiting as a Pre-Return Transnational Strategy Among Youthful Trinidadian Returnees." *Mobilities*, 4, No. 2 (2009): 249-273.

Cornebise, Michael. "Compismo Popular: The Evolution of National Camping Facilities During Cuba's 'Special Period'." *Caribbean Geography*, 11, No. 2 (2000): 81-90.

Cornebise, Michael. "The Social Construction of Tourism in Cuba: A Geographic Analysis of the Representation of Gender and Race During the Special Period." Ph.D. Dissertation, University of Tennessee, Knoxville, Tenn., 2003.

Crist, Raymond E. "Static and Emerging Cultural Landscapes on the Islands of St. Kitts and Nevis, British West Indies." *Economic Geography*, 25, No. 2 (1949): 134-145.

Crist, Raymond E. "Cultural Dichotomy in the Island of Hispanola." *Economic Geography*, 28, No. 2 (1952): 105-121.

Crist, Raymond E. "Changing Cultural Landscapes in Antigua, British West Indies." *American Journal of Economics and Sociology*, 13, No. 3 (1955): 225-231.

Croes, Haime, and Pieter Hooimeijer. "Gender and Chain Migration: The Case of Aruba." *Population, Space, and Place*, 16, No. 2 (2010): 121-134.

Crumplin, William W. "Shelter Supply, Government, and Geography: Housing in Trinidad and Tobago During the Independence Years." Ph.D. Dissertation, Queen's University, Kingston, Ontario, 1994.

Cumberbatch, Carol R. "Isolation of Factors Related to Fertility and Family Planning in Trinidad and Tobago." Master's Thesis, University of Manitoba, Winnipeg, Manitoba, 1981.

Cumper, George. "Population Movements in Jamaica, 1830-1950." *Social and Economic Studies*, 5 (1956): 261-280.

Cundall, Frank. *Jamaican Place Names*. Kingston, Jamaica: The Institute of Jamaica, 1909. 9 pp.

Dacey, M.F. "An Empirical Study of the Areal Distribution of Houses in Puerto Rico." *Transactions, Institute of British Geographers*, 45 (1968): 51-70.

Dale, Edmund H. "The Demographic Problem of the British West Indies." *Scottish Geographical Magazine*, 79 (1963): 23-31.

Dann, Graham M.S., and Robert B. Potter. *Tourism and Postmodernity in a Caribbean Setting*. Aix-en-Provence, France: Universite de Droit, D'Economie et des Sciences, Cahiers du Tourisme, Series C, No. 185, 1994.

Dann, Graham M.S., and Robert B. Potter. "Tourism in Barbados: Rejuvenation or Decline?" Pp. 205-228 in *Island Tourism: Trends and Prospects*, edited by Douglas G. Lockhart and David Drakakis-Smith. London, England: Pinter, 1997.

Dann, Graham, M.S., and Robert B. Potter. "Supplanting the Planters: New Plantations for Old in Barbados." *International Journal of Tourism and Hospitality Research*, 2 (2001): 51-84.

Davis, Elyssa. "Community Participation in Improved Rural Water Supply in the Dominican Republic." Master's Thesis, Hunter College, New York, N.Y., 2009.

Davis, Jeffrey et al. "Military Pollution and Natural Purity: Seeing Nature and Knowing Contamination in Vieques, Puerto Rico." *GeoJournal*, 69, No. 3 (2007): 165-179.

Day, R.J. "Fertility and Female Status: Barbados, 1960-1970." Master's Thesis, University of Windsor, Windsor, Ontario, 1974.

Debysingh, Molly. "Indian Migration: A Study of Settlement and Culture Patterns in Mauritius, the West Indies, and Fiji." Master's Thesis, Syracuse University, Syracuse, N.Y., 1966.

Dehoorne, O., and H. Cao. "Mobility and Tourist Locations: The Case of Martinique Island, a French Territory in the Caribbean." *Belgeo*, 1-2 (2005): 121-134.

Dehoorne, O. et al. "El ecoturismo en el centro de las estrategias de desarrollo. Elements de reflexion a partir de experiencias caribenas." *Revista Geografica*, 148 (2010): 117-136.

Delawarde, Jean B. *La vie paysanne a la Martinque: Essai de geographie humaine*. Fort-de-France, Martinque: Imprimerie officelle, 1937.

Desse, Michel. "Pression anthropique et degradation des littoraux haitiens: L'exemple du golfe de la Gonave." *Les Cahiers d'Outre Mer*, 55, No. 219 (2002): 325-344.

Desse, Michel. "Les difficulties de gestion d'un littoral de service a Haiti: L'exemple du golfe de la Gonave." *Cahiers de Geographie du Quebec*, 47, No. 130 (2003): 63-83.

Desse, Michel. "Systeme d'acteurs et gestion littorale." *Etudes caribeennes*, 2 (2005): 17-31.

Desse, Michel. "Perception et pratiques territoriales des littoraux de la Caraibe." *Etudes caribeennes*, 3 (2005): 35-43.

Desse, Michel. "La plage: Raffet des fractures sociales et ethniques aux Antilles." *Etudes caribeennes*, 4 (2006): 15-27.

Deutschman, Z. "Malaria in North America and the Caribbean Region." *Geographical Review*, 33 (1943): 489-491.

Dittmer, Jason. "Ian Fleming's Jamaica: Spaces of Legitimation and the Bond-age of Popular Culture." *Caribbean Geography*, 15, No. 1 (2008): 14-34.

Dodman, David. "A Place or a People? Social and Cultural Geographies of the Anglophone Caribbean." *Social and Cultural Geographies*, 8, No. 1 (2007): 143-150.

Dodman, David. "Globalization, Tourism, and Local Living Conditions on Jamaica's North Coast." *Singapore Journal of Tropical Geography*, 30, No. 2 (2009): 204-219.

Doerr, Arthur H. "The Relationship of Human Activities in Southwestern Puerto Rico to the Semiarid Climate." Ph.D. Dissertation, Northwestern University, Evanston, Ill., 1951.

Doran, Edwin, Jr. "The West Indian Hip-Roofed Cottage." *California Geographer*, 3 (1962): 97-104.

Dornan, D'Arcy J. "Post-Colonial Linkages Between Tourism and Agriculture in Martinique." Ph.D. Dissertation, University of California, Davis, Calif., 2004.

Duval, David T. "Alternative Tourism on St. Vincent." *Caribbean Geography*, 9, No. 1 (2000): 44-57.

Dyer, Donald R. "Distribution of Population in Hispanola." *Economic Geography*, 30 (1954): 337-346.

Dyson, Alice. "Population Trends in the Eastern Caribbean." Pp. 380-405 in *Liverpool Essays in Geography*, edited by Robert W. Steel and R. Lawton. London, England: Longman, 1967.

Ebanks, G.E. "Differential Internal Migration in Jamaica." *Social and Economic Studies*, 17 (1968): 197-214.

Edward, Carol J. "The Older Generation: An Ecological Study of Old Age in Havana." Ph.D. Dissertation, Syracuse University, Syracuse, N.Y., 1979.

Elliott, Sheryl, and Lisa Neirotti. "Challenges of Tourism in a Dynamic Island Destination: The Case of Cuba." *Tourism Geographies*, 10, No. 3 (2008): 275-402.

Ellis, Mark, Dennis Conway, and Adrian J. Bailey. "The Circular Migration of Puerto Rican Women: Towards a Gendered Explanation." *International Migration Review*, 34, No. 1 (1996): 31-62.

Eyre, L. Alan. "Geographic Aspects of Population Dynamics in Jamaica." Ph.D. Dissertation, University of Maryland, College Park, Md., 1969.

Eyre, L. Alan. *The Shanty Town: A Reappraisal*. Mona, Jamaica: University of the West Indies, Department of Geography Research Note No. 3, 1970.

Eyre, L. Alan. *Geographic Aspects of Population Dynamics in Jamaica*. Boca Raton, Fla.: Florida Atlantic University Press, 1972. 172 pp.

Eyre, L. Alan. "The Ghettoization of an Island Paradise." *Journal of Geography*, 82, No. 5 (1983): 236-239.

Eyre, L. Alan. "Biblical Symbolism and the Role of Fantasy Geography Among the Rastafarians of Jamaica." *Journal of Geography*, 84, No. 4 (1985): 144-148.

Eyre, L. Alan. *Depletion of the Jamaican Rain Forest: The Social and Cultural Consequences*. Quito, Ecuador: Universidad San Francisco de Quito and the Association of Academic Programs in Latin America and the Caribbean, 1992.

Eyre, L. Alan. "The Cockpit Country: A World Heritage Site?" Pp. 259-270 in *Environment and Development in the Caribbean: Geographical Perspectives*, edited by David Barker and Duncan McGregor. Kingston, Jamaica: University of the West Indies Press, 1995.

Eyre, L. Alan. "Ecotourism in Cayman Brac." *Caribbean Geography*, 11, No. 1 (2000): 44-57.

Farmer, Paul. *AIDS and Accusation: Haiti and the Geography of Blame*. Berkeley, Calif.: University of California Press, 1992. 338 pp.

Faulkner, Ella E. "Through the Eyes of a Dougla: An Afro-Indian Re-Examines the Southern Trinidad of Her Childhood." Master's Thesis, University of Oklahoma, Norman, Okla., 2006.

Ferguson, Theresa. "Caribbean Children's Geographies: A Case Study of Jamaica." *Geography Compass*, 1, No. 1 (2007): 46-64.

Fewkes, J. Walter. "Relations of Aboriginal Culture and Environment in the Lesser Antilles." *Bulletin of the American Geographical Society*, 46, No. 9 (1914): 662-678.

Finn, John. "Glocalization: Connecting Globalization to Local Cuban Musicians and Their Music." Master's Thesis, University of Missouri, Columbia, Mo., 2006.

Finn, John. "Contesting Culture: A Case Study of Commodification in Cuban Music." *GeoJournal*, 74, No. 3 (2009): 191-200.

Fisher, Albert L. "Cultural Fragmentation in the Caribbean." *Rocky Mountain Social Science Journal*, 4, No. 1 (1967): 38-48.

Fonaroff, L. Schuyler. *Biogeographic Aspects of Malaria in Trinidad.* Berkeley, Calif.: University of California, Department of Geography, 1966. 67 pp.

Fonaroff, L. Schuyler. "Geographic Notes on the Barbados Malaria Epidemic." *Professional Geographer*, 18, No. 3 (1966): 155-163.

Fonaroff, L. Schuyler. "Man and Malaria in Trinidad: Ecological Perspectives of a Changing Health Hazard." *Annals of the Association of American Geographers*, 58, No. 3 (1968): 526-556.

Fonaroff, L. Schuyler. "The Decline of Malaria in Trinidad." Pp. 165-172 in *Medical Geography: Techniques and Field Studies*, edited by Neil D. McGlashan. London, England: Methuen, 1972.

France, Lesly. "Sustainability and Development in Tourism on the Islands of Barbados, St. Lucia, and Dominica." Pp. 109-125 in *Resource Sustainability and Caribbean Development*, edited by Duncan McGregor et al. Mona, Jamaica: University of West Indies Press, 1998.

France, Lesly, and Brian Wheeler. "Sustainable Tourism in the Caribbean." Pp. 59-69 in *Environment and Development in the Caribbean: Geographical Perspectives*, edited by David Barker and Duncan McGregor. Kingston, Jamaica: University of the West Indies Press, 1995.

Fredrich, Barbara E. "Dooryard Medicinal Plants of St. Lucia." *Yearbook, Association of Pacific Coast Geographers*, 40 (1978): 65-78.

Fritsch, C. "Cruise and Eco-Tourism: Partners or Rivals? The Case of Dominica." Pp. 63-70 in *Land, Sea, and Human Effort in the Caribbean*, edited by B. Ratter and W. Sahr. Hamburg, Germany: Beitrage zur Geographischen Regionalforschung in Lateinamerika, Volume 10, 1997.

Fusco, Audrey C. "Local Food, Sustainability, and Cuba's National Food Program." Master's Thesis, University of Kansas, Lawrence, Kans., 2008.

Gallaway, Lowell F., and Richard K. Vedder. "Location Decisions of Puerto Rican Immigrants to the United States." *Social and Economic Studies*, 20, No. 1 (1971): 188-197.

Giacottino, J.C. "A propos d'une enquete d'antropologie dans un village Martiniquais." *Les Cahiers d'Outre Mer*, 21 (1968): 430-433.

Goelsch, J. *Der Haitianische Vodookult. Eine religionsgeographische Studie*. Salzburg, Austria: Salzburger Geographische Arbeiten, Band 19, 1989.

Gould, Peter, and Anne Lyew-Ayee. "Jamaican Television: Images and Geographic Connections." *Caribbean Geography*, 1, No. 1 (1983): 36-50.

Gould, Peter, and Anne Lyew-Ayee. "Television in the Third World: A High Wind on Jamaica." Pp. 33-62 in *Geography, the Media, and Popular Culture*, edited by J. Burgess and J.R. Gold. London, England: Croom Helm, 1985.

Grossman, Lawrence S. "Diet, Income, and Agriculture in an Eastern Caribbean Village." *Human Ecology*, 26, No. 1 (1998): 21-42.

Guano, Emanuela. "Jamaican Spirit Work: A Performative Approach to Kunins and Revived Zion." Master's Thesis, Louisiana State University, Baton Rouge, La., 1993.

Gulick, Luther. "Rural Occupance in Utuado and Jaguda Municipios, Puerto Rico." Ph.D. Dissertation, University of Chicago, Chicago, Ill., 1952.

Gulick, Luther. "Occupance Summary of Rural Utuado and Jayuga." Pp. 369-402 in *Symposium on the Geography of Puerto Rico*, edited by Clarence Jones and Rafael Pico. Rios Piedras, Puerto Rico: University of Puerto Rico Press, 1955.

Gulick, Luther. "A Puerto Rican Coffee Hacienda." *Oriental Geographer*, 4 (1960): 118-126.

Gullick, Charles. "Afro-American Identity: The Jamaican Nexus." *Journal of Geography*, 82, No. 5 (1983): 205-211.

Gunter, Mario. *Kriterian und Indikatoren als Instrumentarium nachhaltinger Entwicklung. Eine Untersuchung sozialer Nachhaltigkeit am*

Beispiel von Interessengruppen der Forsthewirthschaftung auf Trinidad. Heidelberg, Germany: Heidelberger Geographische Arbeiten, Volume 115, 2002.

Haddock, Kenneth C. "Control of Schistosomiasis: The Puerto Rican Experience." *Social Science and Medicine*, 15D, No. 4 (1981): 501-514.

Haider, Wolfgang. "The Choice Behavior of Winter Beach Vacationers to the Caribbean: An Experimental Approach." Ph.D. Dissertation, McGill University, Montreal, Quebec, 1991.

Haider, Wolfgang, and G.O. Ewing. "A Model of Tourist Choices of Hypothetical Caribbean Destinations." *Leisure Sciences*, 12 (1992): 33-47.

Hall, Andrea. "(Eco)Tourism in the Karst Landscape of Puerto Rico." Master's Thesis, University of Wisconsin, Milwaukee, Wis., 2010.

Halloran, Shirley A. "The Migration of Human Capital from the Third World: A Caribbean Case Study." *Proceedings, Association of American Geographers*, 8 (1976): 50-54.

Hanna, Lewis. "Environmental Hazards in Barbados: Perception and Response in the Scotland District and the West Coast." *Caribbean Geography*, 7, No. 1 (1996): 23-35.

Harrison, L.C. "Dominica: A Wet Tropical Human Habitat." *Economic Geography*, 11, No. 1 (1935): 62-76.

Hawkins, Michael J. "Tourism and Place in Treasure Beach, Jamaica: Imagining Paradise and the Alternative." Ph.D. Dissertation, Louisiana State University, Baton Rouge, La., 1999.

Heffington, Douglas. "Charcoal Making: An Element of Antiguan Folk Culture." *Geographical Bulletin*, 38, No. 2 (1996): 100-108.

Henshall, Janet D. "The Demographic Factor in the Structure of Agriculture in Barbados." *Transactions, Institute of British Geographers*, 38 (1966): 183-196.

Henshall, Janet D. "Post-Emancipation Rural Settlement in the Lesser Antilles." *Proceedings, Association of American Geographers*, 8 (1976): 37-41.

Henshall, Janet D. "Second Homes in the Caribbean." Pp. 75-84 in *Second Homes: Curse or Blessing*, edited by J.T. Coppock. Oxford, England: Pergamon Press, 1977.

Henshall, Janet D. "Women and Small-Scale Farming in the Caribbean." *Conference of Latin Americanist Geographers, Special Publication*, 1 (1981): 44-56.

Hernandez-Alvarez, Jose. "The Movement and Settlement of Puerto Rican Migrants within the United States, 1950-1960." *International Migration Review*, 2 (1963): 40-52.

Hernandez-Alvarez, Jose. "Migration, Return and Development in Puerto Rico." *Economic Development and Cultural Change*, 16 (1966): 574-587.

Hernandex-Alvarez, Jose. *Return Migration to Puerto Rico*. Berkeley, Calif.: University of California Press, 1967. 153 pp.

Higman, Barry W., and Brian J. Hudson. *Jamaican Place Names*. Mona, Jamaica: University of West Indies Press, 2009.

Hills, Theo L. "Land Settlement Schemes: Lessons from the British Caribbean." *Revista Geografica*, 63 (1965): 67-82.

Hills, Theo L., and Jan Lundgren. "The Impact of Tourism in the Caribbean: A Methodological Study." *Annals of Tourism Research*, 4, No. 5 (1977): 248-267.

Hobson, Alexis N. "Mitigation, Preparedness, and Response to the Hurricane Threat: The Case of Nevis in the Aftermath of Hurricane Hugo." Master's Thesis, University of Manitoba, Winnipeg, Manitoba, 1996.

Hoetink, H. "Enkele sociaal-geographische kenmerken van het Caribische Gebied." *Geographische Tijdschrift*, 14 (1961): 145-156.

Honychurch, Lennox. "Caribs, Creoles, and Concepts of Territory: The Boundary Between France and Dominica." *Caribbean Geography*, 6, No. 1 (1995): 61-70.

Horner, Randi. "The Meaning of Home: Barbados from Afar in the Novels of George Lemming and Paule Marshall." Master's Thesis, East Carolina University, Greenville, N.C., 1991.

Horst, Oscar H., and Katsuhiro Asagiri. "The Odyssey of Japanese Colonists in the Dominican Republic." *Geographical Review*, 90, No. 3 (2000): 335-358.

Howell, Christopher D. "Tourism in Tortola, British Virgin Islands: Perceptions Toward Land Carrying Capacity." Ph.D. Dissertation, University of Florida, Gainesville, Fla., 1978.

Hoyt, E., and G.T. Hvenegaard. "A Review of Whale Watching and Whaling with Applications for the Caribbean." *Coastal Management*, 30, No. 4 (2002): 381-399.

Hudson, Brian J. "Tourism and Landscape in Jamaica and Grenada." Pp. 46-59 in *Ambiguous Alternatives: Tourism in Small Developing Countries*, edited by S. Britton and W.C. Clarke. Suva, Fiji: University of the South Pacific, 1987.

Hudson, Brian J. "The Landscapes of Cayuna: Jamaica Through the Sense of John Hearne." *Caribbean Geography*, 3, No. 3 (1992): 187-199.

Hudson, Brian J. "Tourist Brochures as Geography Teaching Material: A Grenadian Example." *Caribbean Geography*, 8, No. 1 (1997): 57-61.

Hunter, John M., and Sonia Arbona. "Disease Rate as an Artifact of the Health Case System: Tuberculosis in Puerto Rico." *Social Science and Medicine*, 19, No. 9 (1984): 997-1008.

Hunter, John M., and Sonia Arbona. "Field Testing Along a Disease Gradient: Some Geographical Dimensions of Tuberculosis in Puerto Rico." *Social Science and Medicine*, 21, No. 9 (1985): 1023-1042.

Innes, Frank C. "Indo-Caribbean Diaspora in Context: A Prospectus for Comparative Studies in Involuntary Migrations." *Indio Caribbean Review*, 2 (1995): 159-163.

Innis, Donald Q. "The Human Ecology of Jamaica." Ph.D. Dissertation, University of California, Berkeley, Calif., 1958.

Jackson, Peter. "Social and Spatial Aspects of Puerto Rican Migration." *Singapore Journal of Tropical Geography*, 1, No. 1 (1980): 37-45.

Jackson, Peter. "Migration and Social Change in Puerto Rico." Pp. 193-213 in *Geography and Ethnic Pluralism*, edited by Colin G. Clarke, David Ley, and Ceri Peach. London England: George Allen and Unwin, 1984.

Johnson, Lillian P. "Geographical Aspects of Schistosoma Mansoni in Puerto Rico." Master's Thesis, Syracuse University, Syracuse, N.Y., 1948.

Johnston, Robert W. "Beauty in the Eye of the Beholder: A Case-Study About Nature, Tourism, and Bermuda." Master's Thesis, Carleton University, Ottawa, Ontario, 2008.

Jolivet, V. "Les Haitiens a Santo Domingo: une masse invisible?" *Bulletin de l'Association de Geographes Francais*, 87, No. 3 (2010): 324-335.

Jones, Clarence F. *The Rural Land Classification Program of Puerto Rico*. Evanston, Ill.: Northwestern University, Department of Geography, 1952.

Jonsson, Cristina, and Dwayne Devonish. "Does Nationality, Gender, and Age Affect Travel Motivation? A Case of Visitors to the Caribbean Island of Barbados." *Journal of Travel and Tourism Marketing*, 25, No. 3-4 (2008): 398-408.

Jutla, Rajinder S. "From Boom to Bust: The Legacy of Oil and Its Impact on Tourism in Trinidad and Tobago." *Tourism*, 49, No. 3 (2001): 203-214.

Keech, James W. "Mortality Patterns in Puerto Rico 1939-1941 and 1969-1971: A Study of Spatial Variations in an Island Population." Ph.D. Dissertation, University of Florida, Gainesville, Fla., 1978.

Kermath, Brian M. "Spatial Dynamics of the Formal and Informal Sector Development in a Resort Area: The Case of Sosua, the Dominican Republic." Master's Thesis, Michigan State University, East Lansing, Mich., 1991.

Kermath, Brian M. "Spatial Dynamics of Resorts: Sosua, Dominican Republic." *Annals of Tourism Research*, 19, No. 2 (1992): 173-190.

Kimber, Clarissa. "Spatial Patterning in the Dooryard Gardens of Puerto Rico." *Geographical Review*, 63, No. 1 (1973): 6-26.

Kimber, Clarissa. "Aboriginal and Peasant Cultures of the Caribbean." *Yearbook, Conference of Latin Americanist Geographers*, 17-18 (1990): 153-164.

King, Alexander. "Raumliche Mobilitat zwischen Paysenerie und Weltmarkt im Verflechtungsraum von Cap-Haitien (Haiti, Departement du Nord)." Pp. 147-173 in *Wirtschafts- und sozialraumliche Strukturwandlungen und Interessenconflikte in Lateinamerika: Beitrage zur geographischen Entwicklungsforschung*, edited by Gerd Kohlhepp. Tubingen, Germany: Tubinger Geographische Studien, Volume 142, 2005.

King, Alexander. *Raumliche Mobilitat in Haiti zwischen Paysannerie und Weltmarkt. Wandel der Beziehungen zwischen Land, Stadt, und Ausland unter dem Einfluss der Globalisierung am Beispiel des Ver-*

flechtungsraum von Cap-Haitien. Tubingen, Germany: Tubinger Geographische Studien, Volume 143, 2005.

Kingsbury, Paul. "Jamaican Tourism and the Politics of Enjoyment." *Geoforum*, 36, No. 1 (2005): 113-132.

Kingsbury, Paul. "Sociospatial Sublimation: The Human Resources of Love in Sandals Resorts International, Jamaica." *Annals of the Association of American Geographers*, 101, No. 3 (2011): 650-669.

Klopfer, S., and Gunter Mertins. "Bevorzugte Zielregionen der Kubanischen Binnenwandrung in der 1990er Jahren." Pp. 269-287 in *Kubas Weg aus der Krise: Neurorganisation der Produktion von Guern und Diestleistungen fur den Export*, edited by Gunter Mertins and Helmut Nuhn. Marburg, Germany: Marburger Geographische Schriften, Heft 138, 2001.

Koster, Rhonda. "Resort Evolution in a Narrowly Based Economy of the Pleasure Periphery: A Case Study of Montego Bay, Jamaica." Master's Thesis, University of Regina, Regina, Saskatchewan, 1997.

Lasserre, Guy. "Les Indiens de la Guadeloupe." *Les Cahiers d'Outre Mer*, 6 (1953): 128-158.

Lasserre, Guy. "Presentation de cartes de densite de la population en Guadeloupe." *Bulletin de la Association de Geographes Francais*, 280-281 (1959): 33-48.

Leone, Frederic. "Socio-Economic Implications of Natural Hazards in the Martinique Island (French West Indies): A GIS Spatial Analysis (in French)." *Annales de Geographie*, 111, No. 5-6 (2002): 549-573.

Lewis, Lawrence A. "The Spatial Properties of Population Mobility Within Puerto Rico." *Journal of Tropical Geography*, 29 (1969): 33-38.

Lloyd-Evans, Sally. "Ethnicity and Gender in the Informal Sector in Trinidad: With Particular Reference to Petty-Commodity Trading." Ph.D. Dissertation, Royal Holloway College, University of London, London, England, 1994.

Lloyd-Evans, Sally, and Robert B. Potter. *Gender, Ethnicity, and the Informal Sector in Trinidad*. Aldershot, England: Ashgate, 2002.

Lopez-Marrero, Tania. "Adaptive Capacity and Resilience to Floods in the Caribbean: A Case Study of Two Flood-Prone Communities in Puerto Rico." Master's Thesis, Pennsylvania State University, University Park, Pa., 2008.

Lopez-Marrero, Tania. "La capracidad de adaptacion a peligros naturales des da una perspective social-comunitaria." *Revista de Ciencias Sociales*, 20 (2009): 106-127.

Lopez-Marrero, Tania. "An Integrative Approach to Study and Promote Natural Hazards Adaptive Capacity." *Geographical Journal*, 176, No. 2 (2010): 150-163.

Lopez-Marrero, Tania, and P. Tschakert. "From Theory to Practice: Enhancing Resilience to Floods in Exposed Communities." *Environment and Urbanization*, 23 (2011): 229-249.

Lopez-Marrero, Tania, and Brent Yarnal. "Putting Adaptive Capacity into the Context of People's Lives: A Case Study of Two Flood-Prone Communities in Puerto Rico." *Natural Hazards*, 52 (2010): 277-297.

Lorah, Paul. "An Unsustainable Path: Tourism's Vulnerability to Environmental Decline in Antigua." *Caribbean Geography*, 6, No. 1 (1995): 28-29.

Lorah, Paul. "Trustafarians and Rastafarians: What Mountain Towns Can Learn from the Development Paths of High-End Caribbean Resorts." Pp. 69-77 in *Mountain Resort Planning and Development in an Era of Globalization*, edited by Thomas Clark, Alison Gill, and Rudi Hartman. New York, N.Y.: Cognizant Communications, 2006.

Lounsbury, John F. "Rural Settlement Features and Their Association to Agricultural Economies in Agua Buenas, Comerio, Corozal, and Naranjito, Puerto Rico." Ph.D. Dissertation, Northwestern University, Evanston, Ill., 1951.

Lounsbury, John F. "Types and Patterns of Rural Settlements in the Aguas Buenas-Naronjito-Comerio-Corozal Area." Pp. 337-368 in *Symposium on the Geography of Puerto Rico*, edited by Clarence Jones and Rafael Pico. Rio Piedras, Puerto Rico: University of Puerto Rico Press, 1955.

Lowenthal, David. "The Population of Barbados." *Social and Economic Studies*, 6, No. 4 (1957): 445-501.

Lowenthal, David. "The Range and Variation of Caribbean Societies." *Annals, New York Academy of Science*, 83 (1960): 786-795.

Lowenthal, David. "The Social Background of the West Indian Federation." Pp. 63-97 in *The West Indies Federation*. Washington, D.C.: American Geographical Society Research Series, No. 23, 1961.

Lowenthal, David. "Caribbean Views of Caribbean Lands." *Canadian Geographer*, 5, No. 2 (1961): 1-9.

Lowenthal, David. "Race and Color in the West Indies." *Daedalus*, 95 (1967): 580-626.

Lowenthal, David. *West Indian Societies*. London, England: Oxford University Press, 1972. 323 pp.

Lowenthal, David, and Colin G. Clarke. "Common Lands, Common Aims: The Distinctive Barbudan Community." Pp. 142-159 in *Peasants, Plantations, and Rural Communities in the Caribbean*, edited by Malcolm Cross and Arnaud F. Marks. Leiden, Netherlands: Royal Institute of Linguistics and Anthropology, 1979.

Maharaj, B. "Mobility and Tourist Locations: The Case of Martinique Island, A French Territory in the Caribbean." *Belgeo*, 1-2 (2005): 121-134.

Mains, Susan. "Island Village, Tourist World: Representing Cultural Geographies of Jamaica." *Caribbean Geography*, 15, No. 1 (2008): 1-13.

Maklin, Fionnuda E. "Tourism Development in Grenada, West Indies." Master's Thesis, University of Calgary, Calgary, Alberta, 1986.

Marieu, J. "Quelques donnees recentes sur la population cubaine." *Les Cahiers d'Outre Mer*, 27 (1974): 128-167.

May, J.M., and D.L. McLellan. *The Ecology of Malnutrition in the Caribbean*. New York, N.Y.: Hofner, 1973. 490 pp.

McElroy, Jerome L., and Klaus de Albuquerque. "The Migration Transition in Small Northern and Eastern Caribbean States." *International Migration Review*, 22 (1988): 30-58.

McElroy, Jerome L., and Klaus de Albuquerque. "Tourism Styles and Policy Responses in the Open Economy-Closed Environment Context." Pp. 143-164 in *Caribbean Ecology and Economics*, edited by N.P. Girvon and D. Simmons. Barbados: Caribbean Conservation Association, 1991.

McElroy, Jerome L., and Klaus de Albuquerque. *An Integrated Sustainable Ecotourism for Small Caribbean Islands*. Bloomington, Ind.: Indiana Center on Global Change and World Peace, Occasional Paper, No. 8, 1992.

McElroy, Jerome L. et al. "Applying the Tourism Destination Life-Cycle Model to Small Caribbean and Pacific Islands." *World Travel and Tourism Review*, 3 (1993): 236-244.

McElroy, Jerome L., and Klaus de Albuquerque. "Community and NGO Initiatives in Coastal Conservation: Lessons from Antigua and Barbuda." *Caribbean Geography*, 8, No. 1 (1997): 18-31.

McElroy, Jerome L., and Klaus de Albuquerque. "Tourism Penetration Index in Small Caribbean Islands." *Annals of Tourism Research*, 25 (1998): 145-168.

McGlashon, Neil D. *A West Indies Geographic Pathology Survey*. Hobart, Australia: University of Tasmania, Department of Geography, Occasional Paper, No. 12, 1982.

McGlashon, Neil D. "Causas de defuncion en diez paises y territorios de habla Ingles a del Caribe." *Boletin de la Oficina Sanitoria Panamerica*, 93, No. 4 (1982): 288-299.

McGlashon, Neil D. "An Epidemiological Spectrum in the Caribbean." *Caribbean Geography*, 1, No. 4 (1984): 234-246.

Mertins, Gunter. "Kuba. Renaissance des alten Modells oder Jonglieren zwischen sozialmus und Marktwortschaft?" *Geographische Rundschau*, 59, No. 1 (2007): 44-50.

Meyer-Arendt, Klaus J., Richard A. Sambrook, and Brian M. Kermath. "Seaside Resorts in the Dominican Republic: A Typology." *Journal of Geography*, 91, No. 5 (1992): 219-225.

Miah, Sophia. "Improving Standards in Bahamian High Schools: Using Geographic Information System as a Pedagogical Tool." Master's Thesis, Carleton University, Ottawa, Ontario, 2009.

Mills, Elizabeth. "The Transnational Community as an Agent for Caribbean Development: Aid from New York City and Toronto to Carriacou, Grenada." *Southeastern Geographer*, 45, No. 2 (2005): 175-191.

Mills, Frank L. "The Cost of Housing in a Tourist Economy: The U.S. Virgin Islands." Pp. 90-110 in *Environment and Development in the Caribbean: Geographical Perspectives*, edited by David Barker and Duncan McGregor. Kingston, Jamaica: University of the West Indies Press, 1995.

Momsen, Janet H. “Gender Selectivity in Caribbean Migration.” Pp. 73-90 in *Gender and Migration in Developing Countries*, edited by. S. Chant. London, England: Belhaven, 1992.

Momsen, Janet H., ed. *Women and Change in the Caribbean: A Pan-Caribbean Perspective*. Bloomington, Ind.: Indiana University Press, 1993. 308 pp.

Momsen, Janet H. “Gender and Environmental Perception in the Eastern Caribbean.” Pp. 57-70 in *The Development Process in Small Island States*, edited by Douglas G. Lockhart, David Drakakis-Smith, and John Schembri. London, England: Routledge, 1993.

Momsen, Janet H. “Tourism, Gender, and Development in the Caribbean.” Pp. 106-120 in *Tourism: A Gender Analysis*, edited by V. Kinnaird and Derek R. Hall. Chichester, England: John Wiley, 1994.

Momsen, Janet H. “Gender Ideology and Land.” Pp. 115-132 in *Caribbean Portraits: Essays on Gender Ideologies and Identities*. Kingston, Jamaica: Ian Randle Publishers, 1998.

Momsen, Janet H. “Caribbean Tourism and Agriculture: New Linkages in the Global Era?” Pp. 115-133 in *Globalization and Neoliberalism: The Caribbean Context*, edited by Thomas Klak. Lanham, Md.: Rowman & Littlefield, 1998.

Momsen, Janet H. “Gender Ideology and Land.” Pp. 115-132 in *Caribbean Portraits: Essays on Gender Ideologies and Identities*, edited by Christine Barrow. Kingston, Jamaica: Ian Randle, 1998.

Momsen, Janet H. “Caribbean Peasantry: Barbadian Farmers Over Four Decades.” *Southeastern Geographer*, 45, No. 2 (2005): 206-221.

Monk, Janice. “Social Change and Sexual Differences in Puerto Rican Rural Migration.” *Conference of Latin Americanist Geographers, Special Publication,* No. 1, (1981): 28-43.

Monk, Janice, and Charles S. Alexander. “Modernization and Rural Population Movements: Western Puerto Rico.” *Journal of Inter-American Studies and World Affairs*, 21 (1979): 523-550.

Monk, Janice, and Charles S. Alexander. “Migration, Development, and the Gender Division of Labour: Puerto Rico and Margarita Island, Venezuela.” Pp. 167-177 in *Women and Change in the Caribbean*, edited by Janet H. Momsen. Indianapolis, Ind.: Indiana University Press, 1993.

Montestruc, E. "Le geographie medicale de la Martinque." *Les Cahiers d'Outre Mer*, 2 (1949): 54-62.

Moral, P. "La maison rurale en Haite." *Les Cahiers d'Outre Mer*, 10 (1957): 117-130.

Morris, Amy. "Tourism in Jamaica: Development or Dependence?" Master's Thesis, University of Georgia, Athens, Ga., 1996.

Mossler, Mariana I. "Hurricane Simulation Efficacy for Increased Event Preparedness: U.S. Virgin Islands." *Middle States Geographer*, 32 (1999): 78-87.

Mullings, Beverley. "Sides of the Same Coin: Coping and Resistance Strategies Among Jamaican Data-Entry Operators." *Annals of the Association of American Geographers*, 89, No. 2 (1999): 290-311.

Mullings, Beverley. "Insider or Outsiders Both or Neither: Some Dilemmas of Interviewing in a Cross-Cultural Setting." *Geoforum*, 30, No. 4 (1999): 337-350.

Munoz, V. "Defense of Neighborhood: Puerto Rican Women's Sense of Place in the Context of Gentrification and Displacement." Ph.D. Dissertation, Syracuse University, Syracuse, N.Y., 1994.

Myroo, Mychelle. "Poverty Alleviation Through Sustainable Tourism: Livelihood Strategies in Anse la Raye, St. Lucia." *Caribbean Geography*, 14, No. 1 (2005): 1-12.

Naidoo, M.B. "The East Indian in Trinidad: A Study of an Immigrant Community." *Journal of Geography*, 59, No. 4 (1960): 175-180.

Nau, Stephanie. *Lokale Akeure in der Kubanischen Transformation. Reaktionen auf den internationalen Tourismus als Faktor der Offaung. Eine sozialgeographischer Beitrag zur aktuellen Kuba-Forschung aus emischer Perspektive*. Passau, Germany: Passauer Schriften zur Geographie, Volume 25, 2008.

Neblett, Jacklyn C. "Caribbean Tourism: A Case for Appropriate Tourism Planning." Master's Thesis, Carleton University, Ottawa, Ontario, 1990.

Nelson, Velvet. "Representation and Images of People, Place, and Natuare in Grenada's Tourism." *Geografiska Annaler B*, 87B, No. 2 (2005): 131-144.

Nelson, Velvet. "R.I.P. Nature Island: The Threat of a Proposed Oil Refinery on Dominica's Identity." *Social and Cultural Geography*, 11, No. 8 (2010): 903-919.

Nelson, Velvet. "The Landscape Reputation: Tourism and Identity in the Caribbean." *Tijdschrift voor Economische en Sociale Geographie*, 102, No. 2 (2011): 176-187.

Nicolas, Thierry. "L'hypo-insularite, une nouvelle condition insulaire. L'exemple des Antilles francaises." *L'espace Geographique*, 34 (2005): 329-341.

Niddrie, David L. *Land Use and Population in Tobago*. Bude, England: The World Land Use Survey, Regional Monograph, No. 3, 1961.

Niddrie, David L. "The Problems of Population Growth in Puerto Rico." *Journal of Tropical Geography*, 20 (1965): 26-33.

Nielsen, Niels C., and Janne J. Liburd. "Geographical Information and Landscape History in Tourism Communication in the Age of Web 2.0: The Case of the Salt River Bay National Park in St. Croix of the U.S. Virgin Islands." *Journal of Travel and Tourism Marketing*, 25, No. 3-4 (2008): 282-298.

Nietschmann, Bernard. "Ecological Change, Inflation and Migration in the Far Western Caribbean." *Geographical Review*, 69, No. 1 (1979): 1-24.

Nietschmann, Bernard. "The Cultural Context of Sea Turtle Hunting in the Caribbean and Problems Caused by Commercial Exploitation." Pp. 67-80 in *Biology and Conservation of Sea Turtles: Proceedings of the World Conference on Sea Turtle Conservation*, edited by K.A. Bjorndal. Washington, D.C.: Smithsonian Press, 1981.

Nonnenmann, R. "Haiti: Raumliche und gesellschaftliche Disparitaten." *Geographische Rundschau*, 37, No. 6 (1985): 296-305.

Novak, John. "Maintenance of a Birth Control Method: The Effects of Attitude and Distance on First-Time Users in Santiago, Dominican Republic." Ph.D. Dissertation, Pennsylvania State University, University Park, Pa., 1983.

Nurse, Leonard. "Development and Change on the Barbados Leeward Coast: A Study of the Human Impact on the Littoral Environment." Ph.D. Dissertation, McGill University, Montreal, Quebec, 1987.

O'Farrell, Andrea. "Tourism and Agriculture on the North Coast of the Dominican Republic." Master's Thesis, Michigan State University, East Lansing, Mich., 1989.

O'Farrell, Andrea. "Tourism and Agriculture on the North Coast of the Dominican Republic." *Revista Geografica*, 113 (1991): 171-192.

Olschki, Leonardo. "The Columbian Nomenclarture of the Lesser Antilles." *Geographical Review*, 33 (1943): 397-414.

Paget, E. "Land-Use and Settlement in Jamaica." Pp. 181-224 in *Geographical Essays on British Tropical Lands*, edited by Robert W. Steel and Charles A. Fisher. London, England: George Philip and Son, 1956.

Pagney, Francoise. "Croissance demographique et envirronnement: Approche et perspective a la Guadaloupe (Antilles Francaises)." *Les Cahiers d'Outre Mer*, 50 (1998): 3-20.

Pailler, G. "La Vitalite demographique du Departement de la Martinique." *Norois*, 12, No. 46 (1965): 217-223.

Palm, Risa, and Michael Hodgson. *Natural Hazards in Puerto Rico: Attitudes of Homeowners*. Boulder, Colo.: University of Colorado Institute of Behavioral Science, Program on Environment and Behavior, Monograph No. 55, 1993. 151 pp.

Paquette, Romain. "An Analysis of the Concentration and Dispersal of Settlement in Martinique." Master's Thesis, McGill University, Montreal, Quebec, 1965.

Parsons, James J. "English-Speaking Settlement in the Western Caribbean." *Yearbook, Association of Pacific Coast Geographers*, 16 (1954): 3-16.

Parsons, James J. *San Andres and Providenica: English-Speaking Islands in the Western Caribbean*. Berkeley, Calif.: University of California, Publications in Geography, 12, No. 1 (1956): 1-84.

Peach, Ceri. "Factors Affecting the Distribution of West Indians in Great Britain." *Transactions, Institute of British Geographers*, 38 (1966): 161-163.

Peach, Ceri. "West Indian Migration to Britain." *International Migration Review*, 1, No. 2 (1967): 34-45.

Peach, Ceri. *West Indian Migration to Britain: A Social Geography*. London, England: Oxford University Press, 1968.

Pearson, Ross. "The Geography of Recreation on a Tropical Island: Jamaica." *Journal of Geography*, 56, No. 1 (1957): 12-21.

Perez, Martha P. "Reciprocity and a Sense of Place: A Phenomenological Map of Haitian Space." Ph.D. Dissertation, University of Texas, Austin, Tex., 2001.

Petit, Joseph-Ernest. *Geographie medicale de la Martinique*. Paris, France: Faculte medicine de Paris, 1932.

Phillips, Joan, and Robert B. Potter. *The Friendship Pattern of Transnational Migrants: The Case of Bajan-Brits and Their Friendships on "Return" to Barbados*. Reading, England: University of Reading, Geography Paper, No. 24, 2006.

Phillips, Joan, and Robert B. Potter. "Black Skins-White Masks: Postcolonial Reflections on Race, Gender, and Sceond Generation Return Migration to the Caribbean." *Singapore Journal of Tropical Geography*, 27, No. 3 (2006): 309-325.

Phillips, Joan, and Robert B. Potter. "Quality of Life Issues and Second-Generation Migration: The Case of 'Bajan-Brit Returnees.'" *Population, Space, and Place*, 15, No. 3 (2009): 239-251.

Poland, Blake D., S. Martin Taylor, and Michael V. Hayes. "The Ecology of Health Services Utilization in Grenada, West Indies." *Social Science and Medicine*, 30, No. 1 (1989): 13-24.

Potter, Robert B. "Congruence Between Space Preferences and Socio-Demographic Structure in Barbados, West Indies: The Use of Cognitive Studies in Third World Urban Planning and Development." *Geoforum*, 14, No. 3 (1983): 249-266.

Potter, Robert B. "Mental Maps and Spatial Variations in Residential Desirability: A Barbados Case." *Caribbean Geography*, 1, No. 3 (1984): 186-197.

Potter, Robert B. "An Analysis of Housing in Grenada, St. Lucia, and St. Vincent, and the Grenadines." *Caribbean Geography*, 3, No. 2 (1991): 107-125.

Potter, Robert B. *Housing Conditions in Barbados: A Geographical Analysis*. Mona, Jamaica: University of the West Indies, Institute of Social and Economic Research, 1992. 129 pp.

Potter, Robert B. "Caribbean Views on Environment and Development: A Cognitive Perspective." *Caribbean Geography*, 3, No. 4 (1992): 236-245.

Potter, Robert B. "Demographic Change in a Small Island State: St. Vincent and the Grenadines, 1980-1991." *Geography*, 77, No. 4 (1992): 374-376.

Potter, Robert B. *Low-Income Housing and the State in the Eastern Caribbean*. Kingston, Jamaica: University of the West Indies Press, 1995. 79 pp.

Potter, Robert B. "The Geography of Relative Affluence and Poverty in Barbados." *Caribbean Geography*, 10, No. 2 (1999): 79-88.

Potter, Robert B. "Social Conditions in St. Lucia: Aggregate Analysis of the 1991 Census at the Quarter Level." *Journal of Eastern Caribbean Studies*, 24 (1999): 25-55.

Potter, Robert B. "Tales of Two Societies: Young Return Migrants to St. Lucia and Barbados." *Caribbean Geography*, 12, No. 1 (2001): 24-43.

Potter, Robert B. "Narratives of Socio-Cultural Adjustment Among Young Return Migrants to St. Lucia and Barbados." *Caribbean Geography*, 12, No. 2 (2001): 70-89.

Potter, Robert B. "The Socio-Demographic Characteristics of Second Generation Return Migrants to St. Lucia and Barbados." Pp. 27-47 in *The Experience of Return Migration: Caribbean Perspectives*, edited by Robert B. Potter, Dennis Conway, and Joan Phillips. Burlington, Vt.: Ashgate, 2005.

Potter, Robert B. "Citizens of Descent: 'Foreign-Born' and 'Young Returning' Nationals to St. Lucia." *Journal of Eastern Caribbean Studies*, 30 (2005): 1-30.

Potter, Robert B. "Young, Gifted, and Back: Second-Generation Transnational Return Migrants to the Caribbean." *Progress in Development Studies*, 5, No. 3 (2005): 213-236.

Potter, Robert B., and Dennis Conway. *Self-Help Housing, the Poor and the State in the Caribbean*. Knoxville, Tenn.: University of Tennessee Press, 1997. 299 pp.

Potter, Robert B., and Dennis Conway. "Experiencing Return: Societal Contributions, Adaptations, and Frustrations." Pp. 283-287 in *The Experience of Return Migration: Caribbean Perspectives*, edited by Robert B. Potter, Dennis Conway, and Joan Phillips. Burlington, Vt.: Ashgate, 2005.

Potter, Robert B., Dennis Conway, and Joan Phillips, eds. *The Experience of Return Migration: Caribbean Perspectives*. Burlington, Vt.: Ashgate, 2005. 293 pp.

Potter, Robert B., Dennis Conway, and Godfrey St. Bernard. "Transnationalism Personified: Young Returning Trinidadians in Their Own Words." *Tijdschrift voor Economische en Sociale Geographie*, 100, No. 1 (2009): 101-113.

Potter, Robert B., and Graham N.S. Dann. "Globalization, Postmodernity, and Development in the Commonwealth Caribbean." Pp. 103-129 in *Global Change and the Commonwealth*, edited by T.M. Yeung. Hong Kong: Chinese University of Hong Kong, 1996.

Potter, Robert B., M. Lloyd, and J. Jacyno. *A Social Atlas of Barbados: Cartographic and Statistical Analysis*. Kingson, Jamaica: Arawak Publications, 2003.

Potter, Robert B., and Joan Phillips. "The Rejuvenation of Tourism in Barbados 1993-2003: Reflections on the Butler Model." *Geography*, 89 (2004): 240-247.

Potter, Robert B., and Joan Phillips. "Both Black and Symbolically White: The 'Bajan-Brit' Return Migrant as Post-Colonial Hybrid." *Ethnic and Racial Studies*, 29 (2006): 901-927.

Potter, Robert B., and Joan Phillips. "The Past Is Still Right Here in the Present: Second-Generation Bajan-Brit Transnational Migrants' Views on Issues Relating to Race and Colour Class." *Environment and Planning D*, 26, No. 1 (2008): 123-145.

Potter, Robert B. et al. "Racism in a Melting Pot…? Trinidadian Mid-Life Transnational Migrants? Views on Race and Colour-Class on Return to Their Homes of Descent." *Geoforum*, 41, No. 5 (2010): 805-813.

Price, A.G. "White Settlement in Saba Island, Dutch West Indies." *Geographical Review*, 24 (1934): 42-60.

Price, Edward T. "The Redlegs of Barbados." *Yearbook, Association of Pacific Coast Geographers*, 19 (1957): 35-40.

Prorok, Carolyn V. "Hindu Temples in Trinidad: A Cultural Geography of Religious Structures and Ethnic Identity." Master's Thesis, Louisiana State University, Baton Rouge, La., 1988.

Prorok, Carolyn V. "Evolution of the Hindu Temple in Trinidad." *Caribbean Geography*, 3, No. 2 (1991): 73-93.

Prorok, Carolyn V. "Trinidad's D'sugared Landscape." *Journal of Cultural Geography*, 27, No. 3 (2010): 287-305.

Prorok, Carolyn V., and Mohammad Hemmasi. "East Indian Muslims and Their Mosques in Trinidad: A Geography of Religious Structures and the Politics of Ethnic Identity." *Caribbean Geography*, 4, No. 1 (1993): 28-48.

Prorok, Carolyn V., and Clarissa Kimber. "The Hindu Temple Gardens of Trinidad: Cultural Continuity and Change in a Caribbean Landscape." *Pennsylvania Geographer*, 35, No. 2 (1997): 98-135.

Pugh, Jonathan. "Social Transformation and Participatory Planning in St. Lucia." *Area*, 37, No. 4 (2005): 384-392.

Pulsipher, Lydia M. "Our Maroon in the Now-Lost Landscapes of Montserrat." *Geographical Review*, 91, No. 1-2 (2001): 132-142.

Rajballie, Ganga B. "Some Aspects of the Settlement Geography of a Trinidad Community." Master's Thesis, University of Alberta, Edmonton, Alberta, 1966.

Rashford, John. "Jamaica's Settlement Vegetation, Agroecology, and the Origin of Agriculture." *Caribbean Geography*, 5, No. 1 (1994): 32-50.

Reed, Gillian. "Health for All: Women's Navigation of the Barbadian Health Care System." Master's Thesis, University of Minnesota, Minneapolis, Minn., 1996.

Regan, Seann. "Health Care Use Pattern in Dominica: Integrating Ethnomedicine in an Era of Biomedicine." Master's Thesis, Miami University, Oxford, Ohio, 2010.

Reid, Basil. "Tainos Not Arawaks: The Indigenous Peoples of Jamaica and the Greater Antilles." *Caribbean Geography*, 5, No. 1 (1994): 67-71.

Reilly, Matthew J. "The Social Costs of Tourism Development: Jineteros in Havana, Cuba: A View from the Streets." Master's Thesis, University of Georgia, Athens, Ga., 2001.

Richardson, Bonham C. "Migration and Death Ceremonies on St. Kitts and Nevis." *Journal of Cultural Geography*, 1, No. 2 (1981): 1-11.

Richardson, Bonham C. *Caribbean Migrants: Environment and Human Survival on St. Kitts and Nevis*. Knoxville, Tenn.: University of Tennessee Press, 1983. 207 pp.

Richardson, Bonham C., and Joseph L. Scarpaci. "The Quality of Life in the Twentieth-Century Caribbean." Pp. 627-669 in *General History of the Caribbean*, Volume 4, edited by Bridget Brereton. London, England: Macmillan Education, 2004.

Roberts, G.W. "Immigration of Africans into the British Caribbean." *Population Studies*, 7 (1954): 235-262.

Roberts, G.W. "Emigration from the Island of Barbados." *Social and Economic Studies*, 4, No. 3 (1955): 245-288.

Roberts, G.W. *The Population of Jamaica*. Cambridge, England: Cambridge University Press, 1957.

Roberts, G.W. "Assessing Internal Migrations in a Small Country." *Canadian Studies in Population*, 5 (1976): 113-130.

Rocheleau, Dianne et al. "Complex Communities and Emergent Ecologies in the Regional Agroforestry of Zambrona-Chacuey, Dominican Republic." *Ecumene*, 8, No. 4 (2001): 465-492.

Rodman, Joseph. "Return Migration to Grenada: Transnational Developments." Master's Thesis, Indiana University, Bloomington, Ind., 2005.

Roseman, Curtis et al. *Social Problems in a Small Jamaican Town*. Urbana, Ill.: University of Illinois, Department of Geography, Occasional Paper, No. 6, 1973. 82 pp.

Rosenberg, T.J. *Residence, Employment and Mobility of Puerto Ricans in New York City*. Chicago, Ill.: University of Chicago, Department of Geography, Research Paper, No. 151, 1974.

Roth, Linda. "Traditional Knowledge and Management of Subtropical Dry Forest: A Study of Resource Use, Landscape Change, and Biodiversity in a Rural Area of the Dominican Republic." Ph.D. Dissertation, Clark University, Worcester, Mass., 1997.

Rouleau, Lucie. "Sondage a deux degrees sur mosaique de photos aeriennes, pour l'estimation de donnees socio-demographiques a Santiago de las Caballeros, Republique Dominicaine." Master's Thesis, University of Sherbrooke, Sherbrooke, Quebec, 1995.

Russo, G. "Arcipelago cubano. Sviluppor poverta e uguaglianza sociale a Cuba." *Bolletino della Societa Geografica Italiana*, Serie 12, 12, No. 1 (2007): 183-202.

Ryan, Michele A. "Institutional Analysis as an Approach to Tourism Development: The Case for Trinidad and Tobago." Master's Thesis, University of Waterloo, Waterloo, Ontario, 1990.

Sagawe, Thorsten. *Una vision Geografica de la Poblacion Dominicana. Atlas Demografico de la Republica Dominicana.* Munich, Germany: Munchener Studien zur Sozial und Wirtschaftsgeographie, Volume 38, 1992.

Sambrook, Richard A., Brian M. Kermath, and Robert N. Thomas. "Seaside Resort Development in the Dominican Republic." *Journal of Cultural Geography*, 12, No. 2 (1992): 65-76.

Sambrook, Richard A., Brian M. Kermath, and Robert N. Thomas. "Tourism Growth Poles Revisited: A Strategy for Regional Economic Development in the Dominican Republic." *Yearbook, Conference of Latin Americanist Geographers*, 20 (1994): 87-96.

Sambrook, Richard A., Bruce W. Pigozzi, and Robert N. Thomas. "Population Pressure, Deforestation, and Land Degradation: A Case Study from the Dominican Republic." *Professional Geographer*, 51, No. 1 (1999): 25-39.

Sammy, R.V. Allen. "Survival Strategies Among Domestic Pure Peasant Farmers in St. Patrick County, Trinidad." Ph.D. Dissertation, University of Western Ontario, London, Ontario, 1977.

Scarpaci, Joseph L. "The Changing Face of Cuban Socialism: Tourism and Planning in the Post-Soviet Era." *Yearbook, Conference of Latin Americanist Geographers*, 24 (1998): 91-110.

Schlosser, Tina C. "Local Realities and Structural Constraints of Agricultural Health: Pesticide Poisoning of Jamaican Small-Holders." Master's Thesis, Virginia Polytechnic Institute and State University, Blacksburg, Va., 1999.

Sharer, Cyrus. "The Population Growth of the Bahamas Islands." Ph.D. Dissertation, University of Michigan, Ann Arbor, Mich., 1955.

Sharkey, Debra A., and Janet H. Momsen. "Tourism in Dominica: Problems and Prospects." *Caribbean Geographer*, 6, No. 1 (1995): 40-51.

Shaw, Earl B. "The Villages of St. Croix." *Bulletin of the Geographical Society of Philadelphia*, 32 (1934): 10-24.

Shaw, Earl B. "Population Adjustments in Our Virgin Islands." *Economic Geography*, 11 (1935): 267-279.

Sheller, Mimi. "Infrastructure of the Imagined Island: Software, Mobilities, and the Architecture of Caribbean Paradise." *Environment and Planning A*, 41, No. 6 (2009): 1386-1403.

Sheller, Mimi. "The New Caribbean Complexity: Mobility Systems, Tourism, and Spatial Rescaling." *Singapore Journal of Tropical Geography*, 30, No. 2 (2009): 189-203.

Silver, Jennifer. "Fisher Participation in a Study of the Socio-Economic and Cultural Value of Sea Turtles in the Turks and Caicos Islands." Master's Thesis, University of Western Ontario, London, Ontario, 2005.

Singh, S.P. "Amerindian Aborigines of Trinidad: Habitat, Economy, Society." *National Geographical Journal of India*, 32, No. 4 (1986): 306-318.

Skelton, Tracey. "Cultures of Land in the Caribbean: A Contribution to the Debate on Development and Culture." *Journal of Development Research*, 8, No. 2 (1996): 71-92.

Sletto, Bjorn et al. "El Rincon de las Olvidados: Participatory GIS, Experiential Learning, and Critical Pedagogy in Santo Domingo, Dominican Republic." *Journal of Latin American Geography*, 9, No. 3 (2010): 111-135.

Slinger-Friedman, Vanessa. "Ecotourism in Dominica: Studying the Potential for Economic Development, Environmental Protection, and Cultural Conservation." *Island Studies Journal*, 4, No. 1 (2009): 3-24.

Sofer, Michael, and Israel Drori. "Land Settlement Schemes in Jamaica and Fiji: Programs Through Transformation." Pp. 179-204 in *The Development Process in Small Island States*, edited by Douglas G. Lockhart, David Drakakis-Smith, and John Schembri. London, England: Routledge, 1993.

Starkey, Otis P. "How We Live in the West Indies." *Journal of Geography*, 59, No. 3 (1960): 123-126.

Stinner, William F., Klaus de Albuquerque, and Roy S. Bryce-Laporte, eds. *Return Migration and Remittances: Developing a Caribbean Perspective*. Washington, D.C.: Smithsonian Institute, 1982. 322 pp.

Street, John M. "Feral Animals in Hispanola." *Geographical Review*, 52, No. 3 (1962): 400-426.

Street, John M. "Problems of Health and Nutrition in Haiti." *Geographical Review*, 61, No. 4 (1971): 599-600.

Symanski, Richard, and Nancy Burley. "The Jewish Colony of Sosuna." *Annals of the Association of American Geographers*, 63, No. 3 (1973): 366-378.

Tata, Robert J. "Too Poor and Small: Caribbean Mini-States." *Focus*, 29, No. 2 (1978): 1-11.

Tata, Robert J., and Arthur S. Evans. "Racial Separation Versus Social Cohesion: The Case of Trinidad-Tobago." *Revista Geografica*, 104 (1986): 23-32.

Thivierge, Pascale. "Tourist Stakeholders, Networks, and Sustainability: The Case of Tourism Development in the Vinales Valley, Pinar de Rio Province, Cuba." Master's Thesis, McGill University, Montreal, Quebec, 2001.

Thomas, Robert N. et al. "Tourist Carrying Capacity Measures: Crowding Syndrome in the Caribbean." *Professional Geographer*, 57, No. 1 (2005): 13-20.

Thomas-Hope, Elizabeth. "The Establishment of a Migration Tradition: British West Indian Movements to the Hispanic Caribbean in the Century After Emancipation." Pp. 66-81 in *Caribbean Social Relations*, edited by Colin G. Clarke. Liverpool, England: University of Liverpool, Centre for Latin American Studies, Monograph Series, No. 8, 1978.

Thomas-Hope, Elizabeth. "Hopes and Realities in the West Indian Migration to Britain." *Oral History*, 8, No. 1 (1980): 35-42.

Thomas-Hope, Elizabeth. "Regional Systems and the Population Component: The Search for Development Strategy in the Caribbean." Pp. 277-299 in *Regional Development Alternatives: International Perspectives*, edited by A.L. Mabogunje and P.R. Misra. Singapore: Maruzen for the United Nations Center for Regional Development, 1981.

Thomas-Hope, Elizabeth. "Off the Island: Population Migration Among the Caribbean Middle Class." Pp. 39-58 in *White Collar Migrants in the Americas and the Caribbean*, edited by Arnaud F. Marks and Hebe M.C. Vessuri. Leiden, Belgium: Royal Institute of Linguistics and Anthropology, 1983.

Thomas-Hope, Elizabeth. "Return Migration and Its Implication for Caribbean Development." Pp. 157-177 in *Migration and Development in the Caribbean*, edited by Robert A. Pastor. Boulder, Colo.: Westview, 1985.

Thomas-Hope, Elizabeth. "Transients and Settlers: Varieties of Caribbean Migrants and the Socio-Economic Implications of Their Return." *International Migration Review*, 24, No. 3 (1986): 559-572.

Thomas-Hope, Elizabeth. "Caribbean International Movement in Urbanization." Pp. 76-91 in *Urbanization and Geographical Distribution of Population*, edited by Committee for International Cooperation in National Research in Demography. Paris, France: The Committee, 1990.

Thomas-Hope, Elizabeth. *Explanations in Caribbean Migration*. London, England: Macmillan, 1993.

Thomas-Hope, Elizabeth. "Population Mobility and Land Assets in Hill Farming Areas of Jamaica." *Caribbean Geography*, 4, No. 1 (1993): 49-63.

Thomas-Hope, Elizabeth. "Small Island Environments and International Migration: The Caribbean." *Scottish Geographical Magazine*, 109, No. 3 (1993): 142-151.

Thomas-Hope, Elizabeth. "Population Mobility in the West Indies: The Role of Perceptual and Environmental Differentials." Ph.D. Dissertation, Oxford University, Oxford, England, 1993.

Thomas-Hope, Elizabeth. *Caribbean Migration*. Kingston, Jamaica: University of West Indies Press, 2002. 184 pp.

Thomas-Hope, Elizabeth, and R.D. Nutter. "Occupation and Status in the Ideology of Caribbean Return Migration." Pp. 97-119 in *The Impact of International Migration on Third World Development*, edited by R.T. Appleyard. Paris, France: Organization for Economic Cooperation and Development, 1988.

Thompson, Christopher. "The Community-Based Homestay Project: A Case Study in Small-Scale Sustainable Tourism Development in the Commonwealth of Dominica." Master's Thesis, Miami University, Oxford, Ohio, 2010.

Tole, L. "Population and Poverty in Jamaican Deforestation: Integrating Satellite and Household Census Data." *GeoJournal*, 57, No. 4 (2002): 227-247.

Trist, Carolyn. "Recreating Ocean Space: Recreational Comsumption and Representation of the Caribbean Marine Environment." *Professional Geographer*, 51, No. 3 (1999): 376-387.

Tunbridge, John. "Large Heritage Waterfronts on Small Tourist Islands: The Case of the Royal Naval Dockyard, Bermuda." *International Journal of Heritage Studies*, 8, No. 2 (2002): 41-51.

Ulbert V. *Partizipative Gender-Forschung. Umveltprobleme und Strategien der Ressourcennutzung in der Dominikanischen Republik.* Freiberg, Germany: Freiburger Studien zur Geographischen Entwicklungsforschung, Volume 17, 1999.

Vetter, M., and P. Hasdenteufel. "Anthropogene Bodendegradation in den stadtnahen gebieten der Provinz Havanna, Kuba: Ausmass, Hintergrunde, und Problemlosungsansatze." *Geo-oko*, 27, No. 1-2 (2006): 20-34.

Vincent, George A.E. "Sustainable Tourism Development in Grenada, West Indies: Towards a Mode of Analysis." Master's Thesis, McGill University, Montreal, Quebec, 1995.

Volkova, Y.D. "Basic Trends in the Population Geography of Contemporary Cuba." *Soviet Geography*, 10, No. 10 (1969): 612-617.

Wall, Geoffrey, and J.M. Ali. "The Impact of Tourism in Trinidad and Tobago." *Annals of Tourism Research*, 5, No. 1 (1977): 43-50.

Walters, N.E. "Land Settlement Schemes in Jamaica: A Socio-Agricultural Study." Master's Thesis, McGill University, Montreal, Quebec, 1966.

Watson, Mark, and Robert B. Potter. "Housing Conditions, Vernacular Architecture, and State Policy in Barbados." Pp. 30-51 in *Self-Help Housing, the Poor, and the State in the Caribbean*, edited by Robert B. Potter and Dennis Conway. Knoxville, Tenn.: University of Tennsessee Press, 1998.

Watson, Mark, and Robert B. Potter. *Low-Cost Housing in Barbados: Evolution or Social Revolution?* Barbados: University of West Indies Press, 2001.

Weaver, David B. "Tobago: The Spatial Development of a Caribbean Tourist Industry." Master's Thesis, Wilfrid Laurier University, Waterloo, Ontario, 1981.

Weaver, David B. "Tourism and the Functional Transformation of the Antiguan Landscape." Pp. 161-176 in *Spatial Implications of Tour-*

ism, edited by C. Fleischer-Van Rooijen. Groningen, Netherlands: GeoPers, 1992.

Weaver, David B. "Ecotourism in the Small Island Caribbean." *Geo-Journal*, 31, No. 4 (1993): 457-466.

Weaver, David B. "Ecotourism in the Caribbean." Pp. 159-176 in *Ecotourism: A Sustainable Option?*, edited by E. Cater and G. Lowman. Chichester, England: Wiley, 1994.

Weaver, David B. "Alternative Tourism in Montserrat." *International Journal of Tourism Management*, 16, No. 8 (1996): 593-604.

Weaver, David B. "Peripheries of the Periphery: Tourism in Tobago and Barbuda." *Annals of Tourism Research*, 25, No. 2 (1998): 292-313.

Webster, Gerald R. "Cuba: Moving Back to the Future with Tourism." *Journal of Geography*, 91, No. 5 (1992): 226-233.

Wehrhahn, Rainer, and Sonke Widderich. "Tourismas als Entwicklungsfaktor im Kubanischen Transformationsprozess." *Erdkunde*, 54, No. 2 (2000): 93-107.

Weil, Connie. "Medical Geographic Research in Latin America and the Caribbean in the 1980s and Beyond." *Yearbook, Conference of Latin Americanist Geographers*, 17-18 (1990): 223-230.

Welch, Barbara M. "Population Density and Emigration in Dominica." *Geographical Journal*, 134, No. 2 (1968): 227-235.

Welsh, Michael J. "Peer Educators' Geographic Range Effect: An Analysis of an AIDS Intervention Program in the Dominican Republic." Ph.D. Dissertation, University of North Carolina, Chapel Hill, N.C., 1994.

Welsh, Michael J. et al. "Evidence of Diffusion from a Targeted HIV/AIDS Intervention in the Dominican Republic." *Journal of Biosocial Science*, 23 (2001): 107-119.

Wesche, Marjorie A. "A Place Name Geography of the Ceded Islands of the Lesser Antilles." Master's Thesis, University of Florida, Gainesville, Fla., 1970.

Widderich, Sonke. *Die sozial en Auswirkungen des Kubanischen Transformation prozesses*. Kiel, Germany: Kieler Geographische Schriften, Heft 106, 2002.

Wiley, James. "Dominica at the Crossroads: Tourism Within Diversification in the Eastern Caribbean." *Middle States Geographer*, 29 (1996): 39-48.

Wilkinson, P. "Travel Agents as Information Brokers: The Cases of Anguilla and Dominica." *Operational Geographer*, 9, No. 3 (1991): 37-41.

Williams, Diana. "Promoting Environmental and Social Gain Through Ecotourism: Cuban and International Perspectives." *Caribbean Geographer*, 7, No. 1 (1996): 52-61.

Williams, L., and B.J. Walter. "Black Caribs in Two Societies: Differential Acculturation Rates." *Ohio Geographer*, 2 (1974): 42-51.

Williams, Patrick. "Tourism Development and Environmental Attitudes in St. Kitts." Master's Thesis, Florida State University, Tallahassee, Fla., 1995.

Woltman, Kelly, and K. Bruce Newbold. "Of Flights and Flotilla: Assimilation and Race in the Cuban Diaspora." *Professional Geographer*, 61, No. 1 (2009): 70-86.

Wood, Chris. "Kids and Space in the Puerto Rican Highlands." *Geographical Review*, 96, No. 2 (2006): 229-258.

Woodfield, Nancy K. "The Role of Ecotourism in Grenada: A Marketing Ploy or a Step Towards Sustainable Development?" Pp. 148-170 in *Resource Sustainability and Caribbean Development*, edited by Duncan McGregor et al. Mona, Jamaica: University of West Indies Press, 1998.

Woodfield, Nancy K. "The Wreck of the Rhone Marine Park, British Virgin Islands: A Social and Economic Impact Study." *Caribbean Geography*, 10, No. 1 (1999): 37-51.

PART 3

Economic Geography

Adams, John E. "Marine Industries of the St. Vincent Grenadines, West Indies." Ph.D. Dissertation, University of Minnesota, Minneapolis, Minn., 1970.

Adams, John E. "Shore Whaling in St. Vincent, West Indies." *Caribbean Quarterly*, 19, No. 1 (1973): 42-50.

Adams, John E. "Offshore Bank Fishing in the West Indies." *Ecumene*, 10 (1978): 20-26.

Adams, John E. "The Fisheries and Fish Markets of St. Vincent Island, Eastern Caribbean." *Singapore Journal of Tropical Geography*, 6, No. 1 (1985): 1-12.

Akin, Wallace E. "The Dairy Industry of the San Juan Area of Puerto Rico." Ph.D. Dissertation, Northwestern University, Evanston, Ill., 1952.

Akin, Wallace E. "The Dairy Industry of the San Juan Area." Pp. 47-70 in *Symposium on the Geography of Puerto Rico*, edited by Clarence Jones and Rafael Pico. Rio Piedras, Puerto Rico: University of Puerto Rico Press, 1955.

Albuquerque, Klaus de, and Jerome L. McElroy. "Agricultural Resurgence in the United States Virgin Islands." *Caribbean Geography*, 1, No. 2 (1983): 121-132.

Ali, Nizam. "Fresh Fish Distribution in Trinidad and Tobago." Master's Thesis, University of Western Ontario, London, Ontario, 1973.

Ammerl, T. "Kuba. Transformation des Wirtschafts und Gesellschaftssystems." *Handbuch des Geographieunterrichts*, 8, No. 2 (2007): 42-52.

Anderson, Elizabeth. "Cuba's Sugar-Soviet Alliance: Reflections on the Past and Post-Cold War Ramifications." *Proceedings, New England-St. Lawrence Valley Division, Association of American Geographers*, 21 (1991): 71-76.

Anderson, Fred L. "Man Versus Machine in the Sugar Industry in Jamaica." Master's Thesis, Southern Connecticut State University, New London, Conn., 1974.

Anderson, J. "Land Use of Barbados." Master's Thesis, McGill University, Montreal, Quebec, 1960.

Anderson, M.G., and P.E. Kneale. "Empirical Approaches to the Improvement of Road Cut Design, with Special Reference to St. Lucia, West Indies." *Singapore Journal of Tropical Geography*, 6, No. 2 (1985): 91-100.

Anderson, M.G. et al. "Reducing Landslide Risk in Communities: Evidence from the Eastern Caribbean." *Applied Geography*, 31, No. 2 (2011): 590-599.

Andre, David J. "Contemporary Economic Development in the British Eastern Caribbean: The Tourist Industry's Potential for Economic Change." Master's Thesis, University of Georgia, Athens, Ga., 1973.

Antonini, Gustavo. "Peasant Agriculture in North West Dominican Republic." *Journal of Tropical Geography*, 32 (1971): 1-10.

Antonini, Gustavo. "Una evaluacion del cambio en el uso de la tierra y el potencial de los recursos naturales en laderas tropicales: Metodologia y application." Pp. 173-238 in *Memoria de la Academia de Ciencias de la Republica Dominicana*, No. 9. Santo Domingo, Dominican Republic: 1985.

Antonini, Gustavo, Katherine C. Ewel, and Howard M. Tupper. *Population and Energy: A Systems Analysis of Resource Utilization in the Dominican Republic*. Gainesville, Fla.: University of Florida Press, 1975.

Antonini, Gustavo, and Mason York. "Integrated Rural Development and the Role of the University of the Caribbean: The Case of Plan Sierra, Dominican Republic." *Revista Geografica*, 90 (1979): 97-114.

Arbona, Sonia. "Land Use and Water Pollution in Puerto Rico." Ph.D. Dissertation, Michigan State University, East Lansing, Mich., 1991.

Arbona, Sonia, and John M. Hunter. "The Geography of Land Use and Water Quality in Puerto Rico." *Yearbook, Conference of Latin Americanist Geographers*, 19 (1993): 93-111.

Arbona, Sonia, and John M. Hunter. "Economic Development Threatens Groundwater in Puerto Rico: Results of a Field Study." *Journal of Geography*, 94, No. 6 (1995): 558-570.

Arbona, Sonia, and John M. Hunter. "Paradise Lost: An Introduction to the Geography of Water Pollution in Puerto Rico." *Social Science and Medicine*, 40, No. 10 (1995): 1331-1356.

Arlt, M. "Auswirkungen der Dritten Agrarreform auf den Tabakonbau in Kuba." Pp. 240-268 in *Kubas Weg aus der Krise: Neuorganisation der Produktion von Guem und Dienstleistungen fur den Export*, edited by Gunter Mertins and Helmut Nuhn. Marburg, Germany: Marburger Geographische Schriften, Heft 138, 2001.

Aspinall, Richard, and Janet H. Momsen. *Small Scale Agriculture in Barbados, 1987*. Newcastle, England: University of Newcastle upon Tyne, Department of Geography, Seminar Paper No. 42, 1987.

Augelli, John P. "Geography of Agriculture and Settlement in Interior Puerto Rico: Problems of the Upper Coiza Basin." Ph.D. Dissertation, Harvard University, Cambridge, Mass., 1950.

Augelli, John P. "Patterns and Problems of Land Tenure in the Lesser Antilles: Antigua, British West Indies." *Economic Geography*, 29 (1953): 362-367.

Augelli, John P. "Sugar Cane and Tobacco: A Comparison of Agricultural Types in the Highlands of Eastern Puerto Rico." *Economic Geography*, 39, No. 1 (1953): 63-73.

Augelli, John P. "Agricultural Colonization in the Dominican Republic." *Economic Geography*, 38, No. 1 (1962): 15-27.

Auty, R.M. "Technological Change in Plantation Transport Systems: A Caribbean Example, 1930-70." *Journal of Tropical Geography*, 41 (1975): 1-8.

Auty, R.M. "Caribbean Sugar Factory Size and Survival." *Annals of the Association of American Geographers*, 66, No. 1 (1976): 76-88.

Auty, R.M. "Transforming Mineral Enclaves: Caribbean Bauxite in the 1970s." *Tijdschrift voor Economische en Sociale Geographie*, 71, No. 3 (1980): 169-179.

Auty, R.M. "MNC Strategy and Regional Growth Potential of Mining: Caribbean Bauxite's Forward Linkage." Pp. 73-116 in *Natural Resources in Tropical Countries*, edited by J.B. Oboi. Singapore: National University of Singapore Press, 1983.

Auty, R.M. "Multinational Resource Corporations, Nationalization, and Diminished Viability: Caribbean Plantations, Mines, and Oilfields in the 1970s." Pp. 160-187 in *Multinational Corporations and the Third World*, edited by Chris Dixon, David Drakakis-Smith, and H.D. Watts. Boulder, Colo.: Westview, 1986.

Auty, R.M. "Intensified Dependence on a Maturing Mining Sector: The Jamaican Bauxite Levy." *Caribbean Geography*, 3, No. 3 (1992): 143-159.

Auty, R.M. "Jamaica's Protracted Recovery Under Orthodox Policy." Pp. 124-142 in *Sustaining Development in Mineral Economics: The Resource Curse Thesis*, edited by R.M. Auty. London, England: Routledge, 1993.

Bahr, Jurgen et al. "Der wirtschaftliche Wandel in Kuba: Reform oder Transformation?" *Geographische Rundschau*, 49, No. 11 (1997): 624-631.

Bahr, Jurgen, and Gunter Mertins. "Die Auswirkungen von Wirtschaftskrise und Wirtschaftsreforem auf das Wanderungsverhaltem in Kuba." *Erdkunde*, 53, No. 1 (1999): 14-34.

Bailey, Steven W. "Farmer Perception of Land Degradation and Land Quality in the Pindars River Sub-Watershed, Clarendon, Jamaica." *Caribbean Geography*, 13, No. 2 (2003): 130-144.

Bain, Nancy, and Fernando Medina. "Development Policy in Puerto Rico, 1952-1980: An Example of Successful Industrial Decentralization." *Yearbook, Conference of Latin Americanist Geographers*, 15 (1989): 79-86.

Bair, A. *The Barbados Fishing Industry*. Montreal, Quebec: McGill University, Department of Geography, Publication No. 6, 1962.

Barker, David. "A Periphery in Genesis and Exodus: Reflections on Rural-Urban Relations in Jamaica." Pp. 294-322 in *The Geography of*

Urban-Rural Interactions in Developing Countries, edited by Robert B. Potter and Timothy Unwin. London, England: Routledge, 1989.

Barker, David. "Dualism and Disasters on a Tropical Island: Constraints on Agricultural Development in Jamaica." *Tijdschrift voor Economische en Sociale Geographie*, 84, No. 5 (1993): 332-340.

Barker, David. "Yam Farmers on the Edge of Cockpit Country: Aspects of Resource Use and Sustainability." Pp. 357-372 in *Resource Sustainability and Caribbean Development*, edited by Duncan McGregor et al. Mona, Jamaica: University of West Indies Press, 1998.

Barker, David, and Clinton L. Beckford. "Yam Production and the Yam Stick Trade in Jamaica: Integrated Problems for Resource Management." Pp. 57-74 in *Resources, Planning, and Environmental Management in a Changing Caribbean*, edited by David Barker and Duncan McGregor. Kingston, Jamaica: University of West Indies Press, 2003.

Barker, David, and Clinton L. Beckford. "Plastic Yam and Plastic Yam Sticks: Perspectives on Indigenous Technical Knowledge Among Jamaican Farmers." *Tijdschrift voor Economische en Sociale Geographie*, 97, No. 5 (2006): 535-546.

Barker, David, and Clinton L. Beckford. "Agricultural Intensification in Jamaican Small-Scale Farming: Vulnerability, Sustainability, and Global Change." *Caribbean Geography*, 15, No. 2 (2008): 160-170.

Barker, David, and Duncan McGregor, eds. *Resources, Planning, and Environmental Management in a Changing Caribbean*. Kingston, Jamaica: University of West Indies Press, 2003.

Barker, David, and David J. Miller. "Farming on the Fringe: Small-Scale Agriculture on the Edge of the Cockpit Country." Pp. 271-292 in *Environmental and Development in the Caribbean: Geographical Perspectives*, edited by David Barker and Duncan McGregor. Kingston, Jamaica: University of the West Indies Press, 1995.

Barros, J. "Une analyse de l'economie haitienne." *Les Cahiers d'Outre Mer*, 21 (1968): 421-424.

Batchelder, Robert B. "The Subhumid Plain of Northwestern Puerto Rico: A Study in Rural Land Utilization." Ph.D. Dissertation, Northwestern University, Evanston, Ill., 1951.

Batchelder, Robert B. "Land Use Problems in the Minor Commercial Agricultural Region of the Hinterland of Isabela." Pp. 129-154 in *Symposium on the Geography of Puerto Rico*, edited by Clarence Jones and Rafael Pico. Rio Piedras, Puerto Rico: University of Puerto Rico Press, 1955.

Bault, A.E. "Deux exemples d'economie rhumiere aux Antilles: Les departements d'Outre-Mer francaises la Republique Dominicaine." *Les Cahiers d'Outre Mer*, 37 (1984): 159-174.

Beckford, Clinton L. "The Informal Commercial Trade in Yam Sticks in Central Jamaica." *Caribbean Geography*, 11, No. 2 (2000): 91-99.

Beckford, Clinton L. "Decision-Making and Innovation Among Small-Scale Yam Farmers in Central Jamaica: A Dynamic, Pragmatic, and Adaptive Process." *Geographical Journal*, 168, No. 3 (2002): 248-259.

Beckford, Clinton L., and David Barker. "Finding Sustainable Ways of Staking Yams and Sourcing Yam Sticks in Jamaica: An Environmental and Economic Imperative." *Caribbean Geography*, 13, No. 2 (2003): 145-155.

Beckford, Clinton L., and David Barker. "The Role and Value of Local Knowledge in Jamaican Agriculture: Adaptation and Change in Small-Scale Farming." *Geographical Journal*, 173, No. 2 (2007): 118-128.

Beckford, Clinton L., David Barker, and Steven W. Bailey. "Adaptation, Innovation, and Domestic Food Production in Jamaica: Some Examples of Survival Strategies of Small-Scale Farmers." *Singapore Journal of Tropical Geography*, 28, No. 3 (2007): 273-296.

Beckford, Clinton L., and Steven W. Bailey. "Vulnerability, Constraints, and Survival on Small-Scale Farms in Southern St. Elizabeth, Jamaica: Strengthening Local Food Production Systems." Pp. 218-236 in *Global Change and Caribbean Vulnerability: Environment, Economy, and Society at Risk*, edited by Duncan McGregor, David Dodman, and David Barker. Mona, Jamaica: University of West Indies Press, 2009.

Beishlog, George. "Trends in Land Use in Southeast Puerto Rico." Ph.D. Dissertation, University of Maryland, College Park, Md., 1953.

Beishlog, George. "Trends in Land Use in Southeastern Puerto Rico." Pp. 269-296 in *Symposium on the Geography of Puerto Rico*, edited by Clarence Jones and Rafael Pico. Rio Piedras, Puerto Rico: University of Puerto Rico Press, 1955.

Belisle, Francois J. "Hotel Food Supply and Local Food Production in Jamaica." Ph.D. Dissertation, University of Georgia, Athens, Ga., 1980.

Belise, Francois J. "Tourism and Food Production in the Caribbean." *Annals of Tourism Research*, 10 (1983): 497-513.

Belisle, Francois J. "Tourism and Food Imports: The Case of Jamaica." *Economic Development and Cultural Change*, 32 (1984): 819-842.

Belisle, Francois J. "Food Production and Tourism in Jamaica: Obstacles to Increasing Local Food Supplies to Hotels." *Journal of Developing Areas*, 19, No. 1 (1984): 1-20.

Belisle, Francois J. "The Significance and Structure of Hotel Food Supply in Jamaica." *Caribbean Geography*, 1, No. 4 (1984): 219-233.

Belisle, Francois J. "Notas Sobre la Articulacion entre el turismo y la produccion de alimentos en Jamaica." *Revista Geografica*, 99 (1984): 99-107.

Bellet, D. Les *Grandes Antilles. Etude de geographie economique*. Paris, France: E. Guilmoto, 1909.

Benoist, J. "Individualisme et traditions techniques chez les pecheurs Martiniquais." *Les Cahiers d'Outre Mer*, 12 (1959): 265-285.

Benoist, J. "Les relations d'une societe et d'un miliieu geographique. La plantation antillaise." Pp. 493-500 in *Etudes de geographie tropicale offertes a Pierre, Gourou*, n.e. Paris, France: Mounton, 1972.

Berleant-Schiller, Riva. "The Social and Economic Role of Cattle in Barbuda." *Geographical Review*, 67, No. 3 (1977): 299-309.

Berleant-Schiller, Riva. "The Failure of Agricultural Development in Post-Emancipation Barbuda: A Study of Social and Economic Continuity in a West Indian Community." *Boletin de Estudios Latinoamericanos y del Caribe*, 25 (1978): 21-36.

Berleant-Schiller, Riva. "Land Tenure, Physical Environment, and Political Change in Barbuda, West Indies." *Proceedings, Middle States Division, Association of American Geographers*, 12 (1979): 1-4.

Berleant-Schiller, Riva. "Grazing and Gardens in Barbuda." Pp. 73-91 in *The Keeping of Animals*, edited by Riva Berleant-Schiller and E. Shaklin. Totowa, N.J.: Rowman & Littlefield, 1983.

Berleant-Schiller, Riva. "Ecology and Politics in Barbudan Land Tenure." Pp. 116-131 in *Land and Development in the Caribbean,* edited by J. Besson and Janet H. Momsen. London, England: Macmillan, 1987.

Berleant-Schiller, Riva, and Lydia M. Pulsipher. "Subsistence Cultivation in the Caribbean." *Nieuwe West-Indische Gids*, 60 (1986): 1-40.

Besson, J., and Janet H. Momsen, eds. *Land and Development in the Caribbean*. London, England: Macmillan, 1987.

Bisson, M. "The Development of Agriculture in South-West Trinidad: A Study of Specialization in a Multi-Racial Society." Master's Thesis, McGill University, Montreal, Quebec, 1968.

Blenman, Maurice. "The Schooner Trade of the Eastern Caribbean." Master's Thesis, University of Calgary, Calgary, Alberta, 1970.

Blenman, Maurice. "Transportation and Regional Integration in the Caribbean." Ph.D. Dissertation, McGill University, Montreal, Quebec, 1976.

Bloch, P. "Kleine Landeskunde von Puerto Rico." *Geographische Rundschau*, 11 (1959): 152-157.

Bloch, P. "Kleine Landeskunde der Dominikanischen Republik." *Geographische Rundschau*, 12 (1960): 351-359.

Bloch, P. "Wirtschaftsgeographie von Puerto Rico." *Wirtschaftsgeographie*, 5 (1961): 14-22.

Blount, Stanley F. "A Regional Analysis of Changes from 1950 to 1961 in Rural Land Use in the Isabela Irrigation District, Northwestern Puerto Rico." Ph.D. Dissertation, Northwestern University, Evanston, Ill., 1963.

Blue, Sarah A. "Remittances, the Informal Economy, and Socio-Economic Inequality in Contemporary Cuban Society." Ph.D. Dissertation, University of California, Los Angeles, Calif., 2004.

Blume, Helmut. "Die gegen wortigen Wandlungen in der Verbreitung von Gross- und Kleinbetrielan auf den Grossen Antillen." Pp. 75-125 in *Beitrage zur Geographie der Neuenwelt*, edited by Wilhem Lauer and Oskar Schmieder. Kiel, Germany: Schriften des Geographischen Instituts der Universitat Kiel, Volume 20, 1961.

Blume, Helmut. "Der Bauxitbergbau auf Jamaika." *Geographische Rundschau*, 14, No. 6 (1962): 227-235.

Blume, Helmut. "Wirtschaftslandschaften und Agrarsozialstrukture in Westindien." *Geographische Rundschau*, 19 (1967): 449-457.

Blume, Helmut. "Types of Agricultural Regions and Land Tenure in the West Indies." *Revista Geografica*, 67 (1967): 7-20.

Blume, Helmut. "Agrarlandschaft und Agrarreform in Kuba." *Geographische Zeitschrift*, 56 (1968): 1-17.

Blume, Helmut. "Kuba: Die karabische Zuckerinsel." *Geographische Rundschau*, 37, No. 6 (1985): 286-295.

Blume, Helmut. "Kubas Zucherwirtschaft vor und nach der Revolution." *Krisengebiet Mittelamerika* (1985): 243-254.

Blume, Helmut. "Cuban Cane Sugar Production: Environmental Determinants and Yields." *Lateinamerika-Studien*, 23 (1987): 125-137.

Boismery, A. "Les relations commerciales entre Marseille et les Antilles." *Les Cahiers d'Outre Mer*, 17 (1964): 386-413.

Boivin, Jean. "Validation d'un modile regional de previsions des besoins en irrigation en milieu tropicale humide: Le cas de Trinidad et Tobago." Master's Thesis, Universite de Montreal, Montreal, Quebec, 1990.

Bolster, Todd. "Working Amongst Contradiction: Paladones and Cuba's Period of Fragmentation." Master's Thesis, Miami University, Oxford, Ohio, 2001.

Bounds, John H. "Land Use in the Bahamas." Ph.D. Dissertation, University of Tennessee, Knoxville, Tenn., 1966.

Bounds, John H. "Industrialization in the Bahamas." *Revista Geografica*, 77 (1972): 95-113.

Bounds, John H. "Restoration of Mined Land to Farming in Jamaica." *Revista Geografica*, 80 (1978): 167-220.

Bounds, John H. "Land Use on New Providence Island, Bahamas, 1960-1979." *Revista Geografica*, 94 (1981): 123-153.

Boyer, Helen M. "Distribution of Sugar Cane Production in Cuba." *Economic Geography*, 15 (1939): 311-325.

Brierly, John S. "Domestic Food Crop Production, Grenada, West Indies." Master's Thesis, University of Alberta, Edmonton, Alberta, 1968.

Brierly, John S. *Small Farming in Grenada, West Indies*. Winnipeg, Manitoba: University of Manitoba, Department of Geography, Manitoba Geographical Series, No. 8, 1974. 308 pp.

Brierly, John S. "Kitchen Gardens in the West Indies, with a Contemporary Study from Grenada." *Journal of Tropical Geography*, 43 (1976): 30-40.

Brierly, John S. "A Review of Development Strategies and Programmes of the People's Revolutionary Government in Grenada, 1979-1983." *Geographical Journal*, 151, No. 1 (1985): 40-52.

Brierly, John S. "The Agricultural Strategies and Progress of the People's Revolutionary Government in Grenada, 1979-1983." *Yearbook, Conference of Latin Americanist Geographers*, 11 (1985): 57-62.

Brierly, John S. "Land Fragmentation and Land-Use Patterns in Grenada." Pp. 194-209 in *Land and Development in the Caribbean*, edited by J. Besson and Janet H. Momsen. London, England: Macmillan, 1987.

Brierly, John S. "A Retrospective on West Indian Small Farming, with an Update on Grenada." Pp. 63-81 in *Small Farming and Peasant Agriculture in the Caribbean*, edited by John S. Brierly and H. Rubenstein. Winnipeg, Manitoba: University of Manitoba, Department of Geography, Manitoba Studies in Geography, No. 10, 1988.

Brierly, John S. "Kitchen Gardens: A Historical Perspective with Current Insights from Grenada." *Food and Nutrition Bulletin*, 7 (1991): 52-60.

Brierly, John S. "A Study of Land Redistribution and the Demise of Grenada's Estate Farming System, 1940-1988." *Journal of Rural Studies*, 8, No. 1 (1992): 67-84.

Brierly, John S. "An Analysis of Land Use of Kitchen Gardens, Grenada." Pp. 125-130 in *A Reader in Caribbean Geography*, edited by David Barker, Carol Newby, and Michael Morrissey. Kingston, Jamaica: Ian Randle Publishers, 1998.

Brierly, John S. "Changing Farming Systems: From Estate Holdings to Model Farms, Gleanings from Grenada, 1940-92." Pp. 253-272 in *Resource Sustainability and Caribbean Development*, edited by Duncan McGregor et al. Mona, Jamaica: University of West Indies Press, 1998.

Brierly, John S. "The Apparent Inflexibility of Small-Scale Farmers to Changes in Grenada's Political-Economic Environment 1982-1992." Pp. 37-56 in *Resources, Planning, and Environmental Management in a Changing Caribbean*, edited by David Barker and Duncan McGregor. Kingston, Jamaica: University of West Indies Press, 2003.

Brierly, John S., and H. Rubenstein, eds. *Small Farming and Peasant Agriculture in the Caribbean*. Winnipeg, Manitoba: University of Manitoba, Department of Geography, Manitoba Studies in Geography, No. 10, 1988.

Brock, D. "Peasant Agriculture in Barbados: A Case Study of Rural Economy." Master's Thesis, McGill University, Montreal, Quebec, 1964.

Brockmann, Vernon W. "Physical Land Types and Land Utilization in the Caguas-San Lorenzo Region of Puerto Rico." Ph.D. Dissertation, Northwestern University, Evanston, Ill., 1951.

Brockmann, Vernon W. "Land Types and Land Utilization in the Caguas-San Lorenzo Region." Pp. 297-324 in *Symposium on the Geography of Puerto Rico*, edited by Clarence Jones and Rafael Pico. Rio Piedras, Puerto Rico: University of Puerto Rico Press, 1955.

Brothers, Timothy S. "Deforestation in the Dominican Republic: A Village View." *Environmental Conservation*, 24, No. 3 (1997): 213-223.

Brown, Eric T. et al. "Determination of Predevelopment Denudation Rates of an Agricultural Watershed (Cayaguas River, Puerto Rico) Using In-Situ-Produced 10Be in River-Born Quartz." *Earth and Planetary Science Letters*, 160 (1998): 723-728.

Brown, William H., Jr. "Marine Fisheries in the British West Indies." Master's Thesis, University of California, Berkeley, Calif., 1967.

Brunnschweiler, Dieter H. "Land Use in Ciales, Morovis, and Orocovis." Pp. 403-440 in *Symposium on the Geography of Puerto Rico*, edited by Clarence Jones and Rafael Pico. Rio Piedras, Puerto Rico: University of Puerto Rico Press, 1955.

Burac, M. "Les zones franches industrielles en Republique Dominicaine." *Les Cahiers d'Outre Mer*, 48 (1995): 5-20.

Burchfield, William W. "Geography of the Pineapple Industry of Puerto Rico." Ph.D. Dissertation, University of Maryland, College Park, Md., 1953.

Burchfield, William W. "The Pineapple Industry." Pp. 97-128 in *Symposium on the Geography of Puerto Rico*, edited by Clarence Jones and Rafael Pico. Rio Piedras, Puerto Rico: University of Puerto Rico Press, 1955.

Burge, Edward J.J. "The Catch Effect of Spiny Lobster (*Panulirus argus*) on South Caicos, British West Indies." *Geographical Bulletin*, 37, No. 2 (1995): 78-96.

Burne, Maurice. *L'economie rurale dans le centre nord de la Martinique (Marigot, Sainte-Marie, Gros-Morne, et Trinite): Contribution a l'etude de la crise d'une economie de mutation*. Talence, France: Travaux et documents de geographie tropicale, No. 21, 1975. 192 pp.

Buzzanell, Peter J. "Physical Resource Constraints on the Agricultural Potentials of a Subtropical Dry Forest Region in the Dominican Republic." Master's Thesis, University of Florida, Gainesville, Fla., 1973.

Bynoe, P. "Land Utilisation in Caribbean Island States and Its Relevance to Guyana." Pp. 45-54 in *Land, Sea, and Human Effort in the Caribbean*, edited by B. Ratter and W. Sahr. Hamburg, Germany: Beitrage zur Geographischen Regionalforschung in Lateinamerika, Volume 10, 1997.

Camara, R. "Les savanes de la Republique Dominicaine et leur exploitation par le systeme dos 'Hatos.'" *Les Cahiers d'Outre Mer*, 53 (2000): 345-366.

Campbell, David S. "The Geography of the Coffee Industry of Puerto Rico." Ph.D. Dissertation, University of Chicago, Chicago, Ill., 1948.

Campbell, Donovan, David Barker, and Duncan McGregor. "Dealing with Drought: Small Farmers and Environmental Hazards in St. Elizabeth, Jamaica." *Applied Geography*, 31, No. 1 (2011): 146-158.

Campbell, J.C. et al. "An Application of Linear Programming and Geographic Information Systems: Cropland Allocation in Antigua." *Environment and Planning A*, 24, No. 4 (1992): 535-550.

Campbell, Rohan. "The Jamaican Dairy and Beef Industries: Some Recent Spatial, Structural, and Development Trends." *Caribbean Geography*, 8, No. 2 (1997): 107-119.

Carman, Sheila S. "The Effectiveness of a Small, Participatory Water Development Project in the St. Vincent Grenadines, West Indies." Master's Thesis, University of Tennessee, Knoxville, Tenn., 1996.

Carr, Liam M., and William D. Heyman. "Jamaica Bound? Marine Resources and Management at a Crossroads in Antigua and Barbuda." *Geographical Journal*, 175, No. 1 (2009): 17-38.

Cecil, Robert G. "Comparative Study of the Fisheries of Martinique and Barbados." Master's Thesis, McGill University, Montreal, Quebec, 1966.

Cecil, Robert G. "Geographical Characteristics of Fisheries in Selected Southeastern Caribbean Islands." Ph.D. Dissertation, McGill University, Montreal, Quebec, 1973.

Chapman, Keith. "Petrochemicals and Economic Development: The Implications of the Puerto Rico Experience." *Professional Geographer*, 34, No. 4 (1982): 405-415.

Chardon, Jean-Pierre. "Le port de Pointe-a-Pitre." *Les Cahiers d'Outre Mer*, 26 (1973): 54-86.

Chardon, Jean-Pierre. "Le role de Puerto Rico dans les relations maritimes de la Caraife." *Les Cahiers d'Outre Mer*, 28 (1975): 137-161.

Chardon, Roland. "Sugar Plantations in the Dominican Republic." *Geographical Review*, 74, No. 4 (1984): 441-454.

Chen, Zheng Xiang. *Land Use and Sugar Industry in Cuba* (in Chinese). Taipei, Taiwan: Fu ming non ye di, 1954.

Chen Young, Paul. "The Costs of Locating an Apparel Industry in Puerto Rico and Jamaica." *Caribbean Studies*, 8, No. 1 (1968): 3-30.

Chivallon, Christine. "The Contested Existence of a Peasantry in Martinique: Scientific Discourses, Controversies, and Evidence." Pp. 159-174 in *Caribbean Land and Development Revisited*, edited by J. Besson and Janet H. Momsen. New York, N.Y.: Palgrave Macmillan, 2007.

Clark, Jeffrey. "Effects of Land Use Change on Northeastern Puerto Rican Rivers." Ph.D. Dissertation, Johns Hopkins University, Baltimore, Md., 1997.

Clausner, Martin D. *Rural Santo Domingo: Settled, Unsettled and Resettled*. Philadelphia, Pa.: Temple University Press, 1973. 323 pp.

Clayton, Alison, and Robert B. Potter. "Industrial Development and Foreign Direct Investment in Barbados." *Geography*, 81, No. 2 (1996): 176-180.

Cleach, A. "Les progress de l'elevage moderne en Guadeloupe." *Les Cahiers d'Outre Mer*, 27 (1974): 22-76.

Coffay, Carolyn. "The Sinking of a Joint Venture: The Helmes-Burton Act and Cuban Telecommunications." Master's Thesis, University of North Carolina, Chapel Hill, N.C., 1999.

Connell, John. "The Cayman Islands: Economic Growth and Immigration in a British Colony." *Caribbean Geography*, 5, No. 1 (1994): 51-66.

Considine, James L. "Moriculture and the Turtling Industry of Grand Cayman: Man's Response to a Vanishing Resource." Master's Thesis, University of South Carolina, Columbia, S.C., 1973.

Conway, Dennis. "The Commuter Zone as a Relocation Choice of Low Income Migrants Moving in a Step-Wise Pattern to Port-of-Spain, Trinidad." *Caribbean Geography*, 1, No. 2 (1983): 89-106.

Conway, Dennis. "Rethinking the Consequences of Remittances for Eastern Caribbean Development." *Caribbean Geography*, 4, No. 2 (1993): 116-130.

Corfield, George S. "Sponge Industry of the Caribbean Area." *Economic Geography*, 14, No. 2 (1938): 201-206.

Coulter, J.K. "Soil and Land Use Problems in the West Indies." *World Crops*, 15 (1963): 349-354.

Courtney, Dale E. "Problems Associated with Predicting Land Use in Low Latitude Humid Regions: A Case Study of the San Sebastian-Rincan Anca, Puerto Rico." Ph.D. Dissertation, University of Washington, Seattle, Wash., 1959.

Courtney, Dale E. "Land Use Trends in San Sebastian and Moca, Puerto Rico." *Yearbook, Association of Pacific Coast Geographers*, 21 (1959): 22-30.

Credner, W. "Probleme der Landnutszung auf den Grossen Antillen." *Zeitschrift Gesellschaft Erdkunde Berlin* (1940): 287-302.

Credner, W. "Typen der Wirtschafts Landschaft auf den Grossen Antillen." *Petermanns Geographische Mitteilungen*, 89 (1943): 1-23.

Credner, W. "Typen der Wirtschafts Landschaft auf den Grossen Antillen." *Wirtschaftsgeographie* (1969): 329-373

Crist, Raymond E. "Sugar Cane and Coffee in Puerto Rico." *American Journal of Economics and Sociology*, 7, No. 1 (1947): 173-184, 321-337, and 469-474.

Crist, Raymond E. "Sugar Cane and Coffee in Puerto Rico, II." *American Journal of Economics and Sociology*, 7, No. 3 (1948): 321-337.

Crist, Raymond E. "Sugar Cane and Coffee in Puerto Rico, III." *American Journal of Economics and Sociology*, 7, No. 4 (1948): 469-474.

Crist, Raymond E. "Some Notes on Recent Trends in Rice Production in Cuba." *Economic Geography*, 28, No. 2 (1956): 126-131.

Crist, Raymond E., and H. Popenoe. "Jamaica and Martinique: Contrasting Aspects of Folk Agriculture and Non-Folk Agriculture." Pp. 501-506 in *Etudes de geographie tropicale offertes a Pierre Gourou.* Paris, France: Mounton, 1972.

Cruz Baez, Angel D. "Export Agriculture Under Economic Development: A Geographic Analysis of the Decline of Sugar Cane Production in Puerto Rico." Ph.D. Dissertation, University of Wisconsin, Madison, Wis., 1977.

Dambaugh, Luella. "Tobacco Production: Vuella Abajo Region, Cuba." *Journal of Geography*, 55, No. 9 (1956): 442-446.

Danklefsen, Mildred M. "The Sugar Industry of Jamaica." Ph.D. Dissertation, Clark University, Worchester, Mass., 1952.

Davies, Craig A. "Export Processing Zones in the Dominican Republic and Jamaica." Master's Thesis, University of Kansas, Lawrence, Kans., 2001.

Davis, Suzanne. "Rethinking Biodiversity Conservation Effectiveness and Evaluation in the National Protected Areas Systems of Tropical Islands: The Case of Jamaica and the Dominican Republic." Ph.D. Dissertation, University of Waterloo-Wilfrid Laurier University Graduate Program, Waterloo, Ontario, 2010.

Davis-Morrison, Vileitha, and David Barker. "Resource Management, Environmental Knowledge, and Decision-Making in Small Farming Systems in the Rio Grande Valley, Jamaica." *Caribbean Geography*, 8, No. 2 (1997): 96-106.

Davis-Morrison, Vileitha. "The Sustainability of Small-Scale Agricultural Systems in the Millbank Area of the Rio Grande Valley, Portland, Jamaica." Pp. 296-316 in *Resource Sustainability and Caribbean Development*, edited by Duncan McGregor et al. Mona, Jamaica: University of West Indies Press, 1998.

Day, Michael J. "The Karst Lands of Antigua: Their Land Use and Conservation." *Geographical Journal*, 173, No. 2 (2007): 170-185.

Deans, Robert H. "Commodity and Geographical Concentration in the CACM Countries and Implications for Economic Development." *Caribbean Studies*, 16, No. 3-4 (1976): 171-206.

De Belle, S.G. "The Cattle Industry of Martinique: A Study in Agricultural Geography." Master's Thesis, McGill University, Montreal, Quebec, 1966.

DeLeon Alejandro, Jose D. "Economic Impact of the Luiz Munoz Marin International Airport in Puerto Rico as Motivated by Potential Flooding Problems Due to Sea-Level Rise." Master's Thesis, University of Illinois, Urbana, Ill., 2007.

Delgado, Gilberto. "Puerto Rico-United States Air Passenger Movement: A Geographical Analysis." Master's Thesis, University of Chicago, Chicago, Ill., 1963.

Desse, Michel. "La recent transformation des acteurs economiques dans les D.O.M. (l'exemple de la Guadeloupe, Martinique, et Reunion)." *Annales de Geographie*, 106 (1997): 592-611.

Desse, Michel. "L'impact economique de la plaissance a la Martinique." *Annales de Geographie*, 109, No. 3 (2000): 306-316.

Dharda, Karen. "Indentured Labor and Trinidad's Integration into the World Economy." Ph.D. Dissertation, Syracuse University, Syracuse, N.Y., 2000.

Dodman, David, and Clare Newstead. "Neoliberalization and Caribbean Agriculture: Trends and Prospects for Livelihoods and National Economies." *Caribbean Geography*, 15, No. 2 (2008): 91-100.

Doerr, Arthur H. "The Salt Industry of Southwestern Puerto Rico." *Journal of Geography*, 52 (1953): 335-338.

Donohgue, J. "The Geographical and Economic Factors Influencing the Markets of Barbados, West Indies." Master's Thesis, McGill University, Montreal, Quebec, 1965.

Doran, Edwin, Jr. "The Caicos Conch Trade." *Geographical Review*, 48, No. 3 (1958): 388-401.

Douzant-Rosenfeld, Denise. "Aspects de l'approvisinnement alimentaire d'une du Tiers Monde: Santo Domingo (Republique Dominicaine)." *L'Espace Geographique*, 16, No. 1 (1987): 101-114.

Downs, Margaret B. "Major Geographic Factors in the Development of Jamaica." Master's Thesis, Columbia University, New York, N.Y., 1933.

Drummond, Ian, and Terry Marsden. "A Case Study of Unsustainability: The Barbados Sugar Industry." *Geography*, 80, No. 4 (1995): 342-354.

Dumaz, B. "Les maraichers de la Cote sous le vent, Guadeloupe: Agriculteurs ou petits patrons?" *Les Cahiers d'Outre Mer*, 39 (1986): 179-204.

Dunfield, Peter. "Some Effects of a Sugar Factory Byproduct Compost on Root Growth and Mycorrhisal Infection of Sugarcane in Barbados." Master's Thesis, McGill University, Montreal, Quebec, 1991.

Dunlop, W.R. "Queensland and Jamaica: A Comparative Study in Geographical Economics." *Geographical Review*, 16, No. 4 (1926): 548-567.

Dupuis, J. "Les paradoxes de Curacao: A travers les provinces de l'empire Shell." *Les Cahiers d'Outre Mer*, 22, No. 85 (1969): 63-74.

Dyer, Donald R. "The Development of Geographic Survey Techniques for the Rural Land Classification Program of Puerto Rico." Ph.D. Dissertation, Northwestern University, Evanston, Ill., 1950.

Dyer, Donald R. "The Pilot Study: The Development of Survey Techniques for the Rural Land Classification Program." Pp. 1-26 in *Symposium on the Geography of Puerto Rico*, edited by Clarence Jones and Rafael Pico. Rio Piedras, Puerto Rico: University of Puerto Rico Press, 1955.

Dyer, Donald R. "Sugar Regions of Cuba." *Economic Geography*, 32 (1956): 177-184.

Dyer, Donald R. "Cuban Sugar Regions." *Revista Geografica*, 67 (1967): 21-30.

Dyson, Alice. "Land Use in the Maracas-St. Joseph Basin, Trinidad." Pp. 261-276 in *Geographers and the Tropics: Liverpool Essays*, edited by Robert W. Steel and R.M. Prothero. London, England: Longman, 1964.

Edwards, David T. "Protection of the Hillsides Occupied by Small Farmers in Jamaica: Lessons of History and Prospects for Public Initiatives." Pp. 341-356 in *Resource Sustainability and Caribbean Development*, edited by Duncan McGregor et al. Mona, Jamaica: University of West Indies Press, 1998.

Emory, S.T. "Our Middle American Trade." *Economic Geography*, 19 (1943): 405-417.

Enjalbert, H. "Le renaissance economique de la Republique dominicain." *Les Cahiers d'Outre Mer*, 5 (1952): 330-356, 6 (1953): 61-87.

Faulkner, Mark D. "Site Selection for a Hypothetical Nuclear Power Plant in Trinidad: Testing the Applicability of U.S. Regulatory Requirements within the Physical and Human Constraints of a Developing Island." Master's Thesis, University of Oklahoma, Norman, Okla., 2007.

Fennell, Thomas. "Rubber from Haiti: A Battle Against Time." *Journal of Geography*, 43, No. 4 (1944): 153-158.

Fentem, Arlin. *Commercial Geography of Dominica*. Bloomington, Ind.: Indiana University, Department of Geography, Technical Report, No. 5, 1960.

Fentem, Arlin. *Commerical Geography of St. Vincent*. Bloomington, Ind.: Indiana University, Department of Geography, Technical Report, No. 10, 1961.

Fentem, Arlin. *Commercial Geography of Antigua*. Bloomington, Ind.: Office of Naval Research Technical Report, No. 11, 1961.

Fernandez, Maria B. "Rice Production in Consolacion del Sur, Cuba." Master's Thesis, Clark University, Worcester, Mass., 1964.

Ferre, Jean-Francois. *Le canne de sucre: Les industries de sucre et du rhum a la Martinique: evolution contemporaine, 1950-1974*. Talence, France: Travaux et documents de geographie tropicale, No. 24, 1976. 320 pp.

Ferre, Jean-Francois. "L'economie sucriere et rhumiere martiniquaise en peril (1950-1980)." *Les Cahiers d'Outre Mer*, 34 (1981): 321-384.

Fingal, Chanelle. "Fair 'Trade Not Aid': Globalisation, Fair Trade, and the St. Lucian Banana Industry." *Caribbean Geography*, 15, No. 2 (2008): 118-129.

Finkel, Hermon J. "Patterns of Land Tenure in the Leeward and Windward Islands and Their Relevance to Problems of Agricultural Development in the West Indies." *Economic Geography*, 40, No. 2 (1964): 163-172.

Flavien, E. "Capitalisme et sous-developpement en Haiti." Master's Thesis, University of Montreal, Montreal, Quebec, 1973.

Flemming, Polita C. "A Matter of Storage: Water Supply on the Caribbean Island of Nevis." Master's Thesis, Southern Illinois University, Carbondale, Ill., 1978.

Floyd, Barry N. "Rural Land Development in Jamaica." *Area*, 2, No. 1 (1970): 7-10.

Floyd, Barry N. "Agricultural Innovations in Jamaica: The Yallahs Valley Land Authority." *Economic Geography*, 46, No. 1 (1970): 63-77.

Floyd, Barry N. "Planning for Rural Development in Jamaica: Spatial Systems Analysis." *Caribbean Quarterly*, 18, No. 1 (1972): 5-13.

Floyd, Barry N. "Agriculture in Jamaica." *Geography*, 57, No. 1 (1972): 32-36.

Floyd, Barry N. *Small-Scale Agriculture in Trinidad: A Caribbean Case Study in the Problems of Transforming Rural Societies in the Tropics*. Durham, England: University of Durham, Department of Geography, 1977. 69 pp.

Floyd, Barry N. "No Easy Harvest: Policies and Priorities for Agriculture in Jamaica." *Journal of Geography*, 82, No. 5 (1983): 212-221.

Foscue, Edwin J. "The Central Highway of Cuba." *Economic Geography*, 9, No. 4 (1933): 406-412.

Found, William C. "Land Use Patterns and Problems in Selected Land Settlements in South-Centrla Jamaica: A Study in Multiple Regression Analysis." Ph.D. Dissertation, University of Florida, Gainesville, Fla., 1966.

Found, William C. "A Multivariate Analysis of Farm Output in Selected Land-Reform Areas of Jamaica." *Canadian Geographer*, 12 (1968): 41-52.

Found, William C. "Feasibility Analysis for an Irrigation Project: The Case of Clarendon, Jamaica." *Papers and Proceedings of the Applied Geography Conferences*, 6 (1983): 219-224.

Fredrich, Barbara E. "Morphology of Dooryard Gardens: Patterns, Imprints, and Transportation in St. Lucia, West Indies." Ph.D. Dissertation, University of California, Los Angeles, Calif., 1975.

Gaise, Gary L., and Dean M. Hanick. "Caribbean Airline Connectivity and Development." *Caribbean Geography*, 1, No. 4 (1984): 272-282.

Galloway, J.H. "Decline of a Staple: The Caribbean Sugar Industry in the Twentieth Century." Pp. 80-104 in *Competing for the Sugar Bowl*, edited by R. Munting and T. Szmrecsangi. St. Katharinen, Germany: Scripta Mercaturae Verlag, 2000.

Gamble, Douglas W. et al. "Climate Change, Drought, and Jamaican Agriculture: Local Knowledge and the Climate Record." *Annals, Association of American Geographers*, 100, No. 4 (2010): 880-893.

George, V., and C. Warren. "The Kingston Waterfront: Evaluation of a Re-development Scheme." *Caribbean Geography*, 1, No. 3 (1984): 203-206.

Gerling, Walter. *Die Plantagenwirtschaft des Rohrzuckers auf den Grossen Antillen. Ein Beitrag zur Agrargeographie der Tropen*. Wurzburg, Germany: Wurtzburger Geographische Arbeiten, Volume 2, 1954.

Gey, P. "Bauerliche Betriebe und Productionsgenossenschaften in Kuba." *Geographische Rundschau*, 41, No. 1 (1989): 22-26.

Giacottino, J.C. "L'economie trinidadienne." *Les Cahiers d'Outre Mer*, 22 (1969): 113-160.

Girault, Christian. "La Commercializacion de Café en la Republica Dominicana: Un Enfoque Geografico." *Revista Geografica*, 88 (1978): 9-60.

Girault, Christian. "Habitants, Speculateurs, et Exporteurs: Le commerce du Café en Haiti." Ph.D. Dissertation, McGill University, Montreal, Quebec, 1980.

Girault, Christian. "Gulf and Western Republique Dominicaise: De l'enclave sucriere au controle d'un partie de l'economie nationale." *L'Espace Geographique*, 9, No. 3 (1980): 223-229.

Girault, Christian. "Recherches geographie economique aux Antilles: La cas haitien." *L'Espace Geographique*, 14, No. 4 (1985): 285-288.

Gordon, Llewelyn. "A Study of the Factors Which Have Determined the Present Economic Developments of Jamaica." Master's Thesis, McGill University, Montreal, Quebec, 1955.

Gould, Peter R., and Anne Lyew-Ayee. "Television in the Third World: A High Wind on Jamaica." Pp. 33-62 in *Geography and the Media*, edited by J. Burgess and J.R. Gold. New York, N.Y.: St. Martin's, 1985.

Green, Marshall. "Planning Models for Efficient Resource Allocation on Peasant Farms: Some Applications in Northwestern Dominican Republic." Ph.D. Dissertation, University of Florida, Gainesville, Fla., 1974.

Greenstein, Lori A. "Site Quality and Changing Land-Use Patterns in Western Puerto Rico." *Great Plains-Rocky Mountain Geographical Journal*, 9 (1980): 40-48.

Gregory, Anthony. "The Turks Island Salt Trade and Industry." Master's Thesis, University of California, Berkeley, Calif., 1970.

Griffith, Anthony D. "The Diversification of Peasant Agriculture in Barbados: Problems of Directed Change in a Small Economy." Master's Thesis, McGill University, Montreal, Quebec, 1973.

Griffith, Anthony D. "A Geographical Perspective on Scale of Farming in a Developing Economy." *Caribbean Quarterly*, 26, No. 3 (1980): 24-35.

Griffith, Anthony D., and Janet F. Foster. "The Airport West-Coast Highway in Barbados: A Source for School-Based Assessment Activities." *Caribbean Geography*, 4, No. 2 (1993): 138-142.

Grossman, Lawrence S. *The Political Ecology of Bananas: Contract Farming, Peasants, and Agrarian Change in the Eastern Caribbean.* Chapel Hill, N.C.: University of North Carolina Press, 1993.

Grossman, Lawrence S. "The Political Ecology of Banana Exports and Local Food Production in St. Vincent, Eastern Caribbean." *Annals of the Association of American Geographers*, 83, No. 2 (1993): 347-367.

Grossman, Lawrence S. "British Aid and Windward Bananas: The Case of St. Vincent and the Grenadines." *Social and Economic Studies*, 43 (1994): 151-179.

Grossman, Lawrence S. *The Political Economy of Bananas: Contract Farming, Peasants, and Agrarian Change in the Eastern Caribbean.* Chapel Hill, N.C.: University of North Carolina Press, 1998.

Grossman, Lawrence S. "Food Consumption, Export Bananas, and Local Crops in a Vincentian Village." *Caribbean Geography*, 9, No. 1 (1998): 1-13.

Grossman, Lawrence S. "Enduring Constraints on Cattle Production in the Caribbean: Changing Patterns on St. Vincent." *Caribbean Geography*, 11, No. 2 (2000): 67-80.

Guilbe, Carlos. "The Globalization of Retail Activities: The Evolution, Expansion, and Spatial Distribution of North American Chain Stores in San Juan, Puerto Rico, 1945-1998." Ph.D. Dissertation, University of Wisconsin, Milwaukee, Wis., 1999.

Guilbe, Carlos. "Big Boxes, Mega Stores, and Category Killers Overseas: The Emerging Geography of North American Chain Stores (NACS) in Puerto Rico." *Southeastern Geographer*, 49, No. 4 (2009): 326-339.

Gunter, Mario. *Kriterien und Indikataren als instrumentarium nachhaltiger Entwicklung. Eine Untersuchung sozialer Nachhaltigkeit am Beispiel von interessengruppen der Forstbewirtschaftung auf Trinidad.* Heidelberg, Germany: Heidelberger Geographische Arbeiten, Heft 115, 2002.

Haas, Hans-Dieter. "Industrialisierrungs: Probleme in Jamaika." *Erdkunde*, 28, No. 2 (1974): 131-137.

Haas, Hans-Dieter. *Industrialisierungsbestrebungen aud den Westindischen Inseln unter besonderer Berucksichtigung von Jamaika und Trinidad.* Tubingen, Germany: Tubinger Geographische Studien, Volume 68, 1976.

Haas, Hans-Dieter. "The Effectiveness of the Industrial Estate Concept in Developing Nations, Taking the Caribbean Countries as an Example." *Applied Geography and Development*, 17 (1981): 17-28.

Haas, Hans-Dieter, and R. Ernst. "Das Problem Mull in Grosstadten der Dritte Weit: Beispiele aus Ecuador und der Dominikanischen Republik." *Zeitschrift fur Wirtschaftsgeographie*, 34, No. 3-4 (1990): 137-150.

Haas, Hans-Dieter, and Thorsten Sagawe. "Recycling als Entwicklung san satz in Dritte Weltlanden." *Geographische Rundschau*, 42, No. 6 (1990): 314-319.

Henshall, Janet D. "The Spatial Structure of Barbadian Peasant Agriculture." Master's Thesis, McGill University, Montreal, Quebec, 1964.

Henshall, Janet D., and Leslie J. King. "Some Structural Characteristics of Peasant Agriculture in Barbados." *Economic Geography*, 42, No. 1 (1966): 74-84.

Hesler, R.H., and M.A. Bernstein. "Woodfuel Use and Sustainable Development in Haiti." *The Energy Journal*, 13, No. 2 (1992): 129-156.

Hills, Theo L. "Land Settlement Schemes: Lessons from the British Caribbean." *Revista Geografica*, 63 (1965): 67-82.

Hills, Theo L., and Stanley Iton. "A Reassessment of the Traditional in Caribbean Small-Scale Agriculture." *Caribbean Geography*, 1, No. 1 (1983): 24-35.

Hobbs, Darryl. "The Cuban Sugar Restructuring Program (2002-2004): A Case Study of Artemisa, Cuba." Master's Thesis, York University, North York, Ontario, 2005.

Honsch, F. "Herausbildung, Probleme, und Tendenzen der Zuckerwirtschaft Kubas." *Geographische Berichte*, 31, No. 2 (1986): 107-123.

Horst, Oscar H. "The Utility of Palms in the Cultural Landscape of the Dominican Republic." *Principas: Journal of the International Palm Society*, 41, No. 1 (1997): 15-28.

Hoy, Don R. "Agricultural Land Use of Guadeloupe." Ph.D. Dissertation, University of Illinois, Urbana, Ill., 1960.

Hoy, Don R. *Agricultural Land Use of Guadeloupe*. Washington, D.C.: National Research Council, Publication No. 884, National Academy of Sciences, 1961. 90 pp.

Hoy, Don R. "Changing Agricultural Land Use on Guadeloupe, French West Indies." *Annals of the Association of American Geographers*, 52, No. 4 (1962): 441-454.

Hoy, Don R. "The Banana Industry of Guadeloupe, French West Indies." *Social and Economic Studies*, 11 (1962): 260-266.

Hoy, Don R., and J. Fisher. "Primary Production and the Measurement of Agricultural Potential in the Caribbean." *Revista Geografica*, 80 (1974): 71-88.

Hoy, Harry E. "Blue Mountain Coffee of Jamaica." *Economic Geography*, 14 (1938): 409-412.

Hubbard, Raymond. *Some Graph-Theoretic Properties of the Jamaican Road Network*. Mona, Jamaica: University of the West Indies, Department of Geography, Paper No. 5, 1971. 29 pp.

Hudson, Brian J. "Wetland Reclamation in Jamaica." *Caribbean Geography*, 1, No. 2 (1983): 75-88.

Huetz de Lemps, C. "La place des plantations dans l'economie portoricaine." *Les Cahiers d'Outre Mer*, 24 (1971): 329-393.

Huetz de Lemps, C. "Le commerce des exclaves vers Cuba." *Les Cahiers d'Outre Mer*, 35 (1982): 187-189.

Hunt, Kristine K. "The Promise of Land Reform in the Dominican Republic." Master's Thesis, Virginia Polytechnic and State University, Blacksburg, Va., 1992.

Husbands, Winston. "The Genesis of Tourism in Barbados: Further Notes on the Welcoming Society." *Caribbean Geography*, 1, No. 2 (1983): 107-120.

Husbands, Winston. "Pattern, Structure, and Formation of Activity Space in Hinterland Resorts." Ph.D. Dissertation, University of Western Ontario, London, Ontario, 1984.

Husbands, Winston. "Leisure Activity Resources and Activity Space Formation in Periphery Resorts: The Response of Tourists and Residents in Barbados." *Canadian Geographer*, 30, No. 3 (1986): 243-249.

Imus, Harold R. "The Mayaguez Area of Puerto Rico: A Study in Farm Economy Analysis." Ph.D. Dissertation, Northwestern University, Evanston, Ill., 1951.

Imus, Harold R. "The Sugar Cane Economy of the Mayaguez Area." Pp. 155-184 in *Symposium on the Geography of Puerto Rico*, edited by Clarence Jones and Rafael Pico. Rio Piedras, Puerto Rico: University of Puerto Rico Press, 1955.

Imus, Harold R. "The Coffee Industry of the Western Highlands." Pp. 441-472 in *Symposium on the Geography of Puerto Rico*, edited by

Clarence Jones and Rafael Pico. Rio Piedras, Puerto Rico: University of Puerto Rico Press, 1955.

Innis, Donald Q. "The Efficiency of Jamaican Peasant Land Use." *Canadian Geographer*, 5, No. 2 (1961): 19-23.

Innis, Donald Q. "The Economic Geography of Jamaica." *Revue Canadian de Geographie*, 17 (1963): 26-30.

Innis, Donald Q. "Peasant Farming: Banana Intercropping in Jamaica." *Proceedings, Association of American Geographers*, 5 (1973): 111-117.

Innis, Donald Q. "Aspects of Jamaican Post-Industrial Agriculture." *Journal of Geography*, 82, No. 5 (1983): 222-226.

Ishimo, Amani et al. "Population Mobility and the Survival of Small Farming in the Rio Grande Valley, Jamaica." *Geographical Journal*, 172, No. 4 (2006): 318-330.

Iton, Stanley. "The Agronomic Characteristics of Small Scale Jamaican Agriculture." Master's Thesis, McGill University, Montreal, Quebec, 1970.

Jalbert, H. "La renaissance economique de la Republique Dominicaine." *Les Cahiers d'Outre Mer*, 6 (1953): 68-87.

James, Preston E. "Geographic Factors in the Trinidad Coconut Industry." *Economic Geography*, 2, No. 2 (1926): 108-125.

Johnson, Althea. "The Artisonal Fishery of the Black River Lower Morass, Jamaica: A Traditional System of Resource Management." Pp. 390-404 in *Resource Sustainability and Caribbean Development*, edited by Duncan McGregor et al. Mona, Jamaica: University of West Indies Press, 1998.

Jones, Clarence, and Hector H. Berrios. "Report on the Rural Land Classification Program of Puerto Rico, 1949-1952." *Revista Geografica*, 44 (1956): 23-40.

Jones, Clarence et al. "The Rural Land Classification Program of Puerto Rico." *Northwestern University Studies in Geography*, 1 (1952): 35-40.

Jones, H. "Tourism and Development: Recent Barbados Experiences." *Town and Country Planning*, 44 (1976): 351-355.

Jones, Paul. "Spatial Impact of Reynolds Jamaica Mines Ltd. on Employment Opportunity: A Simulation Analysis." Master's Thesis, McMaster University, Hamilton, Ontario, 1969.

Jordan, Flora. "Some Aspects of Land Use in the Federation of the West Indies." Master's Thesis, Syracuse University, Syracuse, N.Y., 1960.

Joyce, Ian T. "The Spiny-Lobster Fishery in Cuba." *Geographical Review*, 87, No. 4 (1997): 484-503.

Joyce, Ian T. "Socialism and the Fishery: A Case Study of the Cuban Shrimp Fishery." *Canadian Geographer*, 43, No. 1 (1999): 53-69.

Kent, R.J. "History and Necessity: The Evolution of Soil Conservation Technology in a Jamaican Farming System." *Geographical Journal*, 168, No. 1 (2002): 48-56.

Keys, Ross D. "Puerto Rico's Changing Economic Geography, 1940-1980." *Bulletin, Association of North Dakota Geographers*, 34 (1984): 10-16.

Kimber, Clarissa. "Dooryard Gardens of Martinique." *Yearbook, Association of Pacific Coast Geographers*, 28 (1966): 97-118.

Kingsbury, Nancy. "Attempting Sustainability: Forest Industries and Forest Management Projects in Dominica." Master's Thesis, York University, North York, Ontario, 1993.

Kingsbury, R.C. *Commercial Geography of the Grenadines*. Bloomington, Ind.: Indiana University, Department of Geography, Technical Report, No. 1, 1960.

Kingsbury, R.C. *Commercial Geography of the British Virgin Islands*. Bloomington, Ind.: Indiana University, Department of Geography, Technical Report, No. 2, 1960.

Kingsbury, R.C. *Commercial Geography of Grenada*. Bloomington, Ind.: Indiana University, Department of Geography, Technical Report, No. 3, 1960.

Kingsbury, R.C. *Commercial Geography of Trinidad and Tobago*. Bloomington, Ind.: Indiana University, Department of Geography, Technical Report, No. 4, 1960.

Klak, Thomas. "Maidenforming the Caribbean: Concerns About Jamaica's Industrial Export Promotion Policy." *Caribbean Geography*, 5, No. 2 (1994): 102-116.

Klak, Thomas. "A Framework for Studying Caribbean Industrial Policy." *Economic Geography*, 71, No. 3 (1995): 297-317.

Klak, Thomas. "Distributional Impacts of the 'Free Zone' Component of Structural Adjustment: The Jamaican Experience." *Growth and Change*, 27, No. 3 (1996): 352-387.

Klak, Thomas. "Globalization, Neoliberalism, and Economic Change in Central America and the Caribbean." Pp. 98-126 in *Latin American Transformed: Globalization and Modernity*, edited by R.G. Wynne and C. Kay. London, England: Arnold, 1998.

Koblinski-Siemens, G. "Zur agrargeographischen Gliederung von Haiti." *Erdkunde*, 8, No. 3 (1954): 194-197.

Kopp, A. "L'agriculture a la Guadeloupe." *Annales de Geographie*, 38 (1939): 480-500.

Koster, Rhonda, and Adrian Seaborne. "Resort Evolution in a Narrowly Based Economy of the Pleasure Periphery: A Case Study of Montego Bay, Jamaica." Pp. 197-224 in *Resources, Planning, and Environmental Management in a Changing Caribbean*, edited by David Barker and Duncan McGregor. Kingston, Jamaica: University of West Indies Press, 2003.

Kruger, D. "Commodity Chains in the Cuban Food Economy." *Die Erde*, 138, No. 2 (2007): 187-212.

Laidig, Kermit M. "Land Use in Southeastern Puerto Rico." Ph.D. Dissertation, University of Nebraska, Lincoln, Neb., 1955.

Lambourdiere, E. "Le port de Point-a-Pitre: Une plate-forme logistique dans l'espace maritime caribbean." *Le Cahiers d'Outre Mer*, 52, No. 208 (1999): 373-396.

Landon, Charles E. "Coffee Growing in the Caribbean." *Bulletin of the Philadelphia Geographical Society*, 29 (1930): 280-297.

Lasserre, Guy. "Une plantation de canne aux Antilles: La sucrerie Beauport (Guadeloupe)." *Les Cahiers d'Outre Mer*, 5 (1952): 297-329.

Lawton, Laura. "The Spatial Development of the Caribbean Cruise Ship Industry." Master's Thesis, University of Western Ontario, London, Ontario, 1986.

Lawton, Laura, and Richard W. Butler. "Cruise Ship Industry: Patterns in the Caribbean 1880-1986." *Tourism Management*, 8, No. 4 (1986): 329-343.

Leemann, A. "Barbados." *Zeitschrift fur Wirtschaftsgeographie*, 14, No. 1 (1970): 17-20.

Leemann, A. "Barbados." *Geographica Helvetica*, 25, No. 3 (1970): 130-135.

Leemann, A. "Curacao." *Geographica Helvetica*, 25, No. 3 (1970): 136-140.

Leemann, A. "Jamaika." *Zeitschrift fur Wirtschaftsgeographie*, 16 (1972): 9-14.

Lewis, Lawrence A., and David R. Reynolds. "The Spatial Structure of Transportation Network, and Socio-Economic Development: The Puerto Rican Example." *Proceedings, Indiana Academy of the Social Sciences*, 2 (1967): 24-52.

Limbiro, Arthur. "Land Use and Land Use Legislation for Crown Lands in St. Vincent." *Caribbean Geographer*, 6, No. 2 (1995): 97-111.

Livingston, R. John. "Conservation and Deforestation in the British West Indies: The Case of St. Vincent." Ph.D. Dissertation, University of Kansas, Lawrence, Kans., 2001.

Lloyd-Evans, Sally, and Robert B. Potter. "The Informal Sector of the Economy of the Commonwealth Caribbean." *Bulletin of Eastern Caribbean Affairs*, 17 (1992): 26-40.

Lloyd-Evans, Sally, Duncan McGregor, and David Barker. "Sustainable Development and the Caribbean: Geographical Perspectives." Pp. 3-25 in *Resource Sustainability and Caribbean Development*, edited by Duncan McGregor et al. Mona, Jamaica: University of West Indies Press, 1998.

Lloyd-Evans, Sally. "Gender, Ethnicity, and Small Business Development in Trinidad: Prospects for Sustainable Job Creation." Pp. 195-213 in *Resource Sustainability and Caribbean Development*, edited by Duncan McGregor et al. Mona, Jamaica: University of West Indies Press, 1998.

Lobeck, E.M. "Haiti: A Brief Survey of Its Past and Present Agricultural Problems." *Journal of Geography*, 53 (1954): 277-290.

Lopez-Marrero, Tania et al. "Urban Expansion and the Loss of Prime Agricultural Lands in Puerto Rico." *Ambio*, 30 (2001): 49-54.

Lounsbury, John F. "Farmsteads in Puerto Rico and Their Interpretive Value." *Geographical Review*, 45 (1955): 347-358.

Lounsbury, John F. "Economic Development in Latin America: Puerto Rico as a Case Study." *Revista Geografica*, 60 (1964): 27-32.

Lowenthal, David. "Economic Tribulations in the Caribbean: A Case Study in the British West Indies." *Inter-American Economic Affairs*, 9 (1955): 67-81.

Lowenthal, David. "Production and Population in Jamaica." *Geographical Review*, 48, No. 4 (1958): 568-571.

Lundelius, Ann. "Water Sources: St. Thomas and St. John, American Virgin Islands." *Geographical Bulletin*, 1 (1970): 31-40.

Lyew-Ayee, Anne. "Communications and Information Flows in the Commonwealth Caribbean: A Force for Integration?" Pp. 376-388 in *Collapsing Space and Time: Geographic Aspects of Communications and Information*, edited by Stanley D. Brunn and Thomas R. Leinbach. New York, N.Y.: Harper Collins, 1991.

Machado, Jose A. "Problems of a One-Crop Economy in Cuba." Master's Thesis, Florida State University, Tallahassee, Fla., 1951.

Magnus, Thomas. "The Conch: An Expandable Fold Food of the Bahamas." Master's Thesis, University of Wisconsin, Madison, Wis., 1969.

Maillard, J.C. "La banane en Jamaique et dans les Windward Islands." *Les Cahiers d'Outre Mer*, 22 (1969): 313-324.

Maillard, J.C. "Un Siecle de Speculation Bananiere dans la Zone Antillaise: Bilan et Perpectives." *Revista Geografica*, 80 (1974): 89-103.

Maillard, J.C. "Une periode cruciale dans l'evolution de la bananeraie antillaise: Les annees 1960-1980." *Revista Geografica*, 99 (1984): 109-120.

Manos, Leah D. "Solid Waste Management in Montserrat, West Indies." Master's Thesis, University of Tennessee, Knoxville, Tenn., 1997.

Martinson, Tom L. "The Caribbean Sugar Bowl: A Case Study of 'Human-Environmental' Interaction." *Journal of Geography*, 91, No. 5 (1992): 211-218.

Mbogna, J.P. "Peasant Agriculture in Barbados: A Sample Study." Master's Thesis, McGill University, Montreal, Quebec, 1961.

McCalla, Robert, Brian Slack, and Claude Comtois. "The Caribbean Basin: Adjusting to Global Trends in Containerization." *Maritime Policy and Management,* 32 (2005): 245-261.

McCalla, Robert. "Container Transshipment at Kingston, Jamaica." *Journal of Transport Geography*, 16, No. 3 (2008): 182-190.

McCalla, Robert. "Site and Situation Factores in Transshipment Ports: The Case of the Caribbean Basin." *Tijdschrift voor Economische en Sociale Geographie*, 99, No. 4 (2008): 440-453.

McCoubrey, Andrew. "Traditional Agroforestry for Sustainable Development in Jamaica: Exploratory Observations." Pp. 314-340 in *Resource Sustainability and Caribbean Development*, edited by Duncan McGregor et al. Mono, Jamaica: University of West Indies Press, 1998.

McEachern, J. "The Yuna River Basin, Dominican Republic: Restraints on Agricultural Development." Master's Thesis, McGill University, Montreal, Quebec, 1986.

McElroy, Jerome L., and Klaus de Albuquerque. "Sustainable Small-Scale Agriculture in Small Caribbean Islands." *Society and Natural Resources*, 3 (1990): 109-129.

McGregor, Duncan. "Soil Erosion, Environmental Change, and Development in the Caribbean: A Deepening Crisis?" Pp. 189-208 in *Environment and Development in the Caribbean: Geographical Perspectives*, edited by David Barker and Duncan McGregor. Kingston: University of the West Indies Press, 1995.

McGregor, Duncan. "Developing an Index of Land Degradation: Some Jamaican Examples." *Caribbean Geography*, 9, No. 2 (2000): 136-147.

McGregor, Duncan, and David Barker. "Land Degradation and Hillside Farming in the Fall River Basin, Jamaica." *Applied Geography*, 11, No. 2 (1991): 143-156.

McGregor, Duncan, and David Barker. "Environment, Resources, and Development: Some Reflections on the Caribbean Research Agenda." Pp. 1-10 in *Resources, Planning, and Environmental Management in a Changing Caribbean*, edited by David Barker and Duncan McGregor. Kingston, Jamaica: University of West Indies Press, 2003.

McGregor, Duncan, David Barker, and Donovan Campbell. "Environmental Change and Caribbean Food Security: Recent Hazard Impacts and Domestic Food Production in Jamaica." Pp. 197-217 in *Global Change and Caribbean Vulnerability: Environment, Econ-*

omy, and Society at Risk, edited by by Duncan McGregor, David Dodman, and David Barker. Mona, Jamaica: University of West Indies Press, 2009.

McGregor, Duncan et al. *Resource Sustainability and Caribbean Development*. Mona, Jamaica: University of West Indies Press, 1998. 408 pp.

McNeil, Lois. "Changes in Small Scale Agriculture: The Role of Governments, Westmoreland Parish, Jamaica." Master's Thesis, McGill University, Montreal, Quebec, 1983.

McPhail, Donald D. "The Cattle Farms of Southern Puerto Rico." Pp. 213-244 in *Symposium on the Geography of Puerto Rico*, edited by Clarence Jones and Rafael Pico. Rio Piedras, Puerto Rico: University of Puerto Rico Press, 1955.

McPhail, Donald D. "Transformation in the Pastoral Land Utilization of Southern Puerto Rico." *Revista Geografica*, 17 (1955): 1-7.

McPhail, Donald D. "Puerto Rican Dairying: A Revolution in Tropical Agriculture." *Geographical Review*, 53, No. 2 (1963): 224-246.

McQuillan, D. Aiden. "Estate Production and Transportation, Dominica." Master's Thesis, University of Alberta, Edmonton, Alberta, 1968.

McQuillan, D. Aiden. "Accessibility and the Development of Export Agriculture in Dominica." *Caribbean Geography*, 1, No. 3 (1984): 149-162.

Meikle, Paulette. "Spatio-Temporal Trends in Root Crop Production and Marketing in Jamaica." *Caribbean Geography*, 3, No. 4 (1992): 223-235.

Meikle, Paulette. "The Changing Patterns of Root Crops Production and Marketing in Jamaica." Master's Thesis, University of West Indies, Mona, Jamaica, 1993.

Meikle, Paulette. "Rural Change and Agricultural Sustainability: The Rio Grande Valley Project, Portland, Jamaica." Pp. 273-295 in *Resource Sustainability and Caribbean Development*, edited by Duncan McGregor. Mona, Jamaica: University of West Indies Press, 1998.

Meikle-Yaur, Paulette. "Globalization of Agriculture: Effects on Social and Natural Systems in Rural Communities of Jamaica." *Caribbean Geography*, 14, No. 1 (2005): 40-54.

Menauge, Jacques. *Planteurs et plantations de banane en Guadeloupe*. Talence, France: Travaux et documents de geographie tropicale, No. 14, 1974. 276 pp.

Mertins, Gunter, and Helmut Nuhn. *Kubas Weg aus der Krise: Neuorganisation der Produktion von Gutern und Dienstlestungen fur den Export*. Marburg, Germany: Marburger Geographischen Gesellschaft, 2001. 288 pp.

Mikus, W. "Wirtschaftsson der zonen Objekte der Industrialisierung und Modernisierung in Entwicklungslander mit Beispielen von der Dominikanischen Republik und der Volksrepublik China." Pp. 573-584 in *Festschrift fur Erdmann Gormsen zume 65*, edited by M. Domros and W. Klaer. Mainz, Germany: Mainzer Geographische Studien, Heft 40, 1994.

Miller, Learie A. "A Preliminary Assessment of the Economic Cost of Land Degradation: The Hermitage Catchment, Jamaica." *Caribbean Geography*, 3, No. 4 (1992): 244-252.

Miller, Learie A. "Perspectives on the Sustainability of Protected Areas in Jamaica." *Caribbean Geography*, 10, No. 1 (1999): 57-62.

Miller, Susan C. "Ecological Processes and Local Uses in National Park Armando Bermudez, Dominican Republic." *Caribbean Geography*, 6, No. 2 (1995): 112-127.

Mills, Elizabeth. "Agriculture and Idle Land in Carriacou, Grenada, with a Case Study of the Village of Harvey Vale." Master's Thesis, University of New Mexico, Albuquerque, N.Mex., 1987.

Mills, Elizabeth. "The Transnational Community as an Agent for Caribbean Development: Aid from New York City and Toronto to Carriacou, Grenada." *Southeastern Geographer*, 45, No. 2 (2005): 175-191.

Mills, Elizabeth. "Leave to Come Back: The Importance of Family Land in a Transnational Caribbean Community." Pp. 233-242 in *Caribbean Land and Development Revisited*, edited by J. Besson and Janet H. Momsen. New York, N.Y.: Palgrave Macmillan, 2007.

Mills, Frank L. "The Rural Structure of Agriculture in St. Kitts." Master's Thesis, University of Western Ontario, London, Ontario, 1971.

Mills, Frank L. "The Development of Alternative Farming Systems and Prospects for Change in the Structure of Agriculture in St. Kitts,

West Indies." Ph.D. Dissertation, Clark University, Worcester, Mass., 1974.

Milstead, Harley P. "The Cacao Industry of Grenada." *Economic Geography*, 36 (1960): 195-203.

Momsen, Janet H. "The Geography of Land Use and Population in the Caribbean (with Special Reference to Barbados and the Windward Islands)." Ph.D. Dissertation, Kings College, University of London, London, England, 1969.

Momsen, Janet H. "Land Settlement as an Imposed Solution." Pp. 46-69 in *Land and Development in the Caribbean*, edited by J. Besson and Janet H. Momsen. London, England: Macmillan, 1987.

Momsen, Janet H. "Development and Gender Divisions of Labor in the Rural Eastern Caribbean." Pp. 232-246 in *Women and Change in the Caribbean: A Pan-Caribbean Perspective*, edited by Janet H. Momsen. Bloomington, Ind.: Indiana University Press, 1993.

Momsen, Janet H. "Linkages Between Tourism and Agriculture in the Caribbean." Pp. 55-61 in *Land, Sea, and Human Effort in the Caribbean*, edited by B. Ratter and W. Sahr. Hamburg, Germany: Beitrage zur Geographischen Regionalforschung in Lateinamerika, Volume 10, 1997.

Momsen, Janet H. "The Waxing and Waning of Land for the Peasantry in Barbados." Pp. 175-188 in *Caribbean Land and Development Revisited*, edited by J. Besson and Janet H. Momsen. New York, N.Y.: Palgrave Macmillan, 2007.

Momsen, Janet H. "Fair Trade Versus Unfair Trade: The Windward Island Banana Industry." *Caribbean Geography*, 15, No. 2 (2008): 130-141.

Montoya, David G., and Hector H. Berrios. "Uses Made of the Data Obtained in the Rural Land Classification Program." Pp. 473-486 in *Symposium on the Geography of Puerto Rico*, edited by Clarence Jones and Rafael Pico. Rio Piedres, Puerto Rico: University of Puerto Rico Press, 1955.

Moral, P. "La culture du café en Haiti." *Les Cahiers d'Outre Mer*, 8 (1955): 233-256.

Morgan, D.J. "The Growth of Jamaica." *Geographical Magazine*, 60 (1967): 588-598.

Morley, D. "Applied Geography or Action Research: The St. Lucia Energy Futures Project." *Geography Research Forum*, 6 (1983): 3-23.

Mulchansingh, V.C. *CARIFTA, New Horizons in the West Indies*. Mona, Jamaica: University of the West Indies, Department of Geography, Occasional Paper, No. 3, 1968.

Mulchansingh, V.C. *Trends in the Industrialization of Jamaica*. Mona, Jamaica: University of the West Indies, Department of Geography, Occasional Paper, No. 6, 1970.

Mulchansingh, V.C. "Multinational Corporations." *Social and Economic Studies*, 19, No. 4 (1970). Special Number.

Mulchansingh, V.C. "The Bauxite/Alumina Industry of Jamaica." *Journal of Tropical Geography*, 33 (1971): 20-30.

Mulchansingh, V.C. "The Location of Oil Refining in Latin America and the Caribbean." *Revista Geografica*, 75 (1971): 85-126.

Mulchansingh, V.C. "The Oil Industry in the Economy of Trinidad." *Caribbean Studies*, 11 (1971): 73-100.

Mulchansingh, V.C. "Extra-Kingston Location of Industry." *Revista Geografica*, 85 (1977): 149-188.

Mulchansingh, V.C. "The Economic Geography of Caribbean Energy." *Proceedings, Conference of Latin Americanist Geographers*, 9 (1983): 144-171.

Mulchansingh, V.C. "Use of Natural Gas in Trinidad and the Point Lisas Pole de Croissance." *Caribbean Geography*, 1, No. 2 (1983): 133-138.

Mullaney, Emma G. "Carib Territory: Indigenous Access to Land in the Commonwealth of Dominica." *Journal of Latin American Geography*, 8, No. 2 (2009): 71-96.

Mullings, Beverley. "Telecommunications Restructuring and the Development of Export Information Processing Services in Jamaica." Pp. 163-184 in *Globalization and Caribbean Identity*, edited by H. Dunn. Kingston, Jamaica: Ian Randle Ltd., 1995.

Mullings, Beverley. "Industrial Development in an Era of Structural Adjustments: The Growth of Export Information Services in Jamaica." Ph.D. Dissertation, McGill University, Montreal, Quebec, 1996.

Mullings, Beverley. "Globalization and the Territorialization of the New Caribbean Service Economy." *Journal of Economic Geography*, 4, No. 3 (2004): 275-298.

Naley, David F. "Land Use in Guanica, Yauco, Guayanilla, and Penuelas." Pp. 245-268 in *Symposium on the Geography of Puerto Rico*, edited by Clarence Jones and Rafael Pico. Rio Piedras, Puerto Rico: University of Puerto Rico Press, 1955.

Nanita-Kennett, Milagros. "Industrial Free Zones in the Dominican Republic." *Caribbean Geography*, 3, No. 3 (1992): 200-204.

Nash, James. "From Periphery to the Core: EPZs in the Caribbean." *Bulletin, Association of North Dakota Geographers*, 45 (1995): 65-78.

Nelson, Erin. "A Better World Is Possible: Agroecology as a Response to Socio-Economic and Political Conditions in Cuba." Master's Thesis, University of Waterloo, Waterloo, Ontario, 2006.

Netzer, Donald L. "Climate and Agricultural Cropland Use in Puerto Rico." Ph.D. Dissertation, University of Illinois, Urbana, Ill., 1953.

Niddrie, David L. *Land Use and Population in Tobago*. Bude, England: World Land Use Survey, Regional Monograph, No. 3, 1961. 59 pp.

Norton, Ann, and Richard Symanski. "The Internal Marketing Systems of Jamaica." *Geographical Review*, 65, No. 4 (1975): 461-475.

Norton, Ann, and Richard Symanski. "The Internal Marketing Systems of Jamaica." *Ekistics*, 42 (1976): 55-61.

Nuhn, Helmut. "Reorientierung des Kubanischen Aussenhandels nach dem Zerfall des COMECON: Perspektiven fur traditionelle und neue Exportproducte." Pp. 36-59 in *Kubas Weg aus der Krise: Neurorganisation der Produktion von Gutern und Dienstleistung fur den Exports*, edited by Gunter Mertins and Helmut Nuhn. Marburg, Germany: Marburger Geographische Schriften, Heft 138, 2001.

Nuhn, Helmut. "Kubas Strategie dur einen biotechnologisch-pharmazeutischen Forchungspal." *Zeitschrift fur Wirtschaftsgeographie*, 47, No. 2 (2003): 122-130.

Odell, P.R. "The Development of the Middle Eastern and Caribbean Refining Industries 1939-63." *Tijdschrift voor Economische en Sociale Geographie*, 54 (1963): 202-213.

Oguntoyinbo, J.S. "Rainfall Variability on Barbados and Its Influence upon Sugar Cane Yield." Master's Thesis, McGill University, Montreal, Quebec, 1964.

Oxtoby, F.E. "The Role of Political Factors in the Virgin Islands Watch Industry." *Geographical Review*, 60, No. 4 (1970): 463-474.

Oyelese, J.O. "The Cultivation of Food Crops in Barbados." Master's Thesis, McGill University, Montreal, Quebec, 1964.

Oyelese, J.O. "Some Aspects of Food Crop Cultivation in the Plantations of Barbados, W.I." *Nigerian Geographical Journal*, 9 (1966): 55-70.

Paget, E. "Value, Valuation, and Use of Land in the West Indies." *Geographical Journal*, 127 (1961): 493-498.

Palmer, Ernest C. "Land Use and Landscape Change Along the Dominican-Haitian Borderlands." Ph.D. Dissertation, University of Florida, Gainesville, Fla., 1976.

Palmer, Jesse. "The Banana in Caribbean Trade." *Economic Geography*, 8 (1932): 262-273.

Pamuk, Ayse, and David E. Dowell. "The Price of Land for Housing in Trinidad: The Role of Regulatory Constraints and Implications for Affordability." *Urban Studies*, 35, No. 2 (1998): 285-300.

Parafina, Sofronio. "Household Relations and Agricultural Decision Making in Haiti." Master's Thesis, University of Florida, Gainesville, Fla., 1993.

Pearson, Ross. "The Jamaican Bauxite Industry." *Journal of Geography*, 56, No. 8 (1957): 377-384.

Perez-Luan, Agapito. "Factors Associated with the Diffusion of Innovations in Rice Cultivation in the Dominican Republic." Ph.D. Dissertation, University of Florida, Gainesville, Fla., 1979.

Pico, Rafael. *Studies in the Economic Geography of Puerto Rico*. Rio Piedras, Puerto Rico: University of Puerto Rico Bulletin, Series 8, No. 1, 1937.

Pico, Rafael. "Land Tenure in the Leasing Types of Farming of Puerto Rico." *Economic Geography*, 15, No. 2 (1939): 135-145.

Pico, Rafael. *Economic Growth in Puerto Rico*. San Juan, Puerto Rico: n.p., 1959.

Piraszewska, K. "Economic Significance of Tourism in the Lesser Antilles." *Miscellanea Geographica*, 12 (2006): 269-276.

Platt, Raye R. "Economic and Social Problems in the British West Indies: A Review." *Geographical Review*, 30, No. 4 (1940): 672-675.

Platt, Robert S. "Resources and Economic Interests of the Bermudas." Ph.D. Dissertation, University of Chicago, Chicago, Ill., 1920.

Platt, Robert S. "A Classification of Manufacturers, Exemplified by Puerto Rican Industries." *Annals of the Association of American Geographers*, 17 (1927): 79-91.

Platt, Robert S. "Geography of a Sugar District, Mariel, Cuba." *Geographical Review*, 19, No. 4 (1929): 603-612.

Pollard, H.J. "Tourism: The Growth Industry in the Caribbean." *Geography*, 61 (1967): 102-107.

Pollard, H.J. "Geographical Variations within the Tourism Trade of the Caribbean." *Journal of Tropical Geography*, 43 (1976): 49-62.

Pollard, H.J. "Planned Animal Husbandry Schemes in Trinidad: The Expectations and the Experience." *Economic Geography*, 51, No. 1 (1980): 18-29.

Pollard, H.J. "Food Crop Production for Trinidad's Home Market: An Unfulfilled Potential." *Singapore Journal of Tropical Geography*, 2, No. 2 (1981): 91-100.

Pollard, H.J., and D.A. Eastwood. "Dairying and Development at Wallerfield, Trinidad." *Tijdschrift voor Economische en Sociale Geographie*, 67, No. 5 (1976): 289-299.

Possekel, Anja K. "Offshore Financial Centres in the Caribbean: Potential and Pitfalls." *Caribbean Geography*, 7, No. 2 (1996): 81-96.

Potter, Amy, and Andrew Sluyter. "Renegotiating Barbuda's Commons: Recent Changes in Barbudan Open-Range Cattle Herding." *Journal of Cultural Geography*, 27, No. 2 (2010): 129-150.

Potter, Robert B. "Industrial Development and Urban Planning in Barbados." *Geography*, 68, (1983): 46-50.

Potter, Robert B., and Graham M.S. Dean. "Some Observations Concerning Postmodernity and Sustainable Development in the Caribbean." *Caribbean Geography*, 5, No. 2 (1994): 92-101.

Power, Dominic, and Daniel Hallencreutz. "Profiting from Creativity? The Music Industry in Stockholm, Sweden, and Kingston, Jamaica." *Environment and Planning A*, 34, No. 10 (2002): 1833-1854.

Quittelier, B. "Bananeraies et environnement : L'impact du mode d'exploitation. Comparison des cas costaricain et martinique." *Belgeo*, 1 (2009): 83-103.

Ramon, Maria. "Recent Development of the Marine Fisheries in Cuba, the Gulf of Bataboro, and the Port of Surgiduro as a Case Study." Master's Thesis, University of California, Berkeley, Calif., 1970.

Ramphall, Davin. "Caribbean Industrialization as Sustainable Development: Exploring the Contradictions." *Caribbean Geography*, 5, No. 2 (1994): 81-91.

Ranely-Verge-Depre, Colette. "Quinze annees de conteneurisation des trafics maritimes aux Antilles franais." *Le Cahiers d'Outre Mer*, 50, No. 198 (1997): 151-170.

Ranely-Verge-Depre, Colette. "Transport aerien et territories insulaires: L'exemple des Petites Antilles." *Annales de Geographie*, 117 (2008): 97-109.

Ranely-Verge Depre, Colette. "Les dynamiques aeroportuaires dans l'archipel des Antilles." *Bulletin de l'Association de Geographes Francais*, 4 (2009): 512-522.

Rapai, George. "Small-Scale Farming in St. Lucia." Master's Thesis, University of Western Ontario, London, Ontario, 1984.

Rappenecker, Caspar. "The Regional and Economic Geography of Jamaica, British West Indies." Ph.D. Dissertation, Cornell University, Ithaca, N.Y., 1936.

Ratter, B. "Complex Resource Management: Requirements for Sustainable Development on Small Caribbean Islands." *Geographische Zeitschrift*, 84, No. 2 (1996): 114-124.

Ratter, B., and A. Droge. "Vom Companero zum Camerero: Des stille Niedergong der Kubanischen Zuckerwirtschaft." *Geographische Rundschau*, 57, No. 9 (2005): 58-63.

Reading, Alison J., and J. Sousson. "Small Farm Production in Dominica, West Indies: A Strategy for Survival." *Land Degradation and Rehabilitation*, 1, No. 2 (1989): 153-168.

Rees, Peter W. "Air Transport in the Lesser Antilles." Master's Thesis, University of California, Berkeley, Calif., 1964.

Reid-Henry, Simon. "Under the Microscrope: Fieldwork Practice and Cuba's Biotechnology Industry: A Reflexive Affair?" *Singapore Journal of Tropical Geography*, 24, No. 2 (2003): 184-197.

Revert, E. "L'economie martiniquaise." *Les Cahiers d'Outre Mer*, 1 (1948): 28-39.

Rhiney, Kevon. "Towards a New Model for Improved Tourism-Agriculture Linkages? The Case of Two Farmers' Co-operatives in Jamaica." *Caribbean Geography*, 15, No. 2 (2008): 142-159.

Richardson, Bonham C. "The Agricultural Dilemma of the Post-Plantation Caribbean." *Inter-American Economic Affairs*, 26 (1972): 59-70.

Richardson, Bonham C. "Labor Migrants from the Island of Carriacou: Workers on the Margins of the Developed World." *Proceedings, Association of American Geographers*, 6 (1974): 149-152.

Richardson-Ngwenya, Pamela. "The EU Sugar Reform and the Responses of Caribbean Sugar Producers." *Geography*, 95, No. 2 (2010): 70-79.

Ridgley, Mark A. et al. "Optimization of Economics, Policies, and Investment Projects Using a Fuzzy-Logic-Based Cost-Effectiveness Model of Coral Reef Quality: Empirical Results from Montego Bay, Jamaica." *Coral Reefs*, 18, No. 4 (1999): 381-392.

Roberts, Susan M. "Small Place, Big Money: The Cayman Islands and the International Financial System." *Economic Geography*, 71, No. 3 (1995): 237-256.

Robinson, William. "Land Tenure in the Leeward and Windward Islands, British West Indies." Master's Thesis, University of Michigan, Ann Arbor, Mich., 1969.

Rocheleau, Dianne, and Laurie Ross. "Trees as Tools, Trees as Text: Struggles Over Resources in Zambrana-Chacuez, Dominican Republic." *Antipode*, 27, No. 4 (1995): 407-428.

Roosevelt, T. "Land Problems in Puerto Rico and the Philippine Islands." *Geographical Review*, 24 (1934): 182-204.

Roy, Harry E. "Blue Mountain Coffee in Jamaica." *Economic Geography*, 14, No. 4 (1938): 409-412.

Royle, Stephen A. "Offshore Finance and Tourism as Development Strategies: Bermuda and the British West Indies." Pp. 126-147 in

Sustainability and Development in the Caribbean: Geographical Perspectives, edited by Duncan McGregor et al. Kingston, Jamaica: University of West Indies Press, 1998.

Ryder, Roy. "Point Score Analysis of Agricultural Decision-Making in the Dominican Republic." *Caribbean Geography*, 4, No. 1 (1993): 2-14.

Ryder, Roy. "Land Evaluation for Steepland Agriculture in the Dominican Republic." *Geographical Journal*, 160, No. 1 (1994): 74-86.

Ryder, Roy. "Farmer Perception of Soils in the Mountains of the Dominican Republic." *Mountain Research and Development*, 14, No. 3 (1994): 261-266.

Ryder, Roy. "La Tecnologia Agricola en la Republica Dominicana." *Revista Geografica*, 120 (1994): 79-90.

Sachtler, Manfred. "Some Results of the Forest Inventory in the Dominican Republic and Their Implications for Future Development of Forestry." *Applied Geography and Development*, 12 (1974): 111-132.

Sagawe, Thorsten. "Mining as an Agent for Regional Development: The Case of the Dominican Republic." *Geography*, 74, No. 1 (1989): 69-71.

Sagawe, Thorsten. "Industrial Free Zones in the Dominican Republic: National Versus Local Impact." *Journal of Geography*, 95, No. 5 (1996): 203-212.

Sammy, R.V. Allen. "The Development of Commercial Sugar Cane Agriculture in Trinidad to 1838." Master's Thesis, University of Alberta, Edmonton, Alberta, 1967.

Scarpaci, Joseph L. "Globalization and Work: Economy, Remittances, and Joint-Ventures in Post-Soviet Cuba." *Transnational Law and Contemporary Problems*, 14 (2004): 101-117.

Schaeper, Herbert R.A. "Technological Adaptations to Solar and Aeolian Energy Utilization: Two Case Studies in Rural Jamaica." Ph.D. Dissertation, University of Florida, Gainesville, Fla., 1969.

Schaeper, Herbert F.A. "Alternative Energy Resource Development in the West Indies." *Papers of the Applied Geography Conferences*, 4 (1981): 364-372.

Scherf, K. "Okonomische-geographische Strukturwandlungen der Republik Kuba." *Geographische Berichte*, 9 (1964): 149-177.

Scott, Calford I. "An Analysis of the Supply Characteristics of Jamaican Sugar Cane Farmers." Master's Thesis, McMaster University, Hamilton, Ontario, 1971.

Semevskiy, B.N. "Basic National-Economic Problems of the Republic of Cuba." *Soviet Geography*, 7, No. 1 (1966): 36-43.

Semevskiy, B.N. "Social and Natural Conditions of Cuban Agriculture." *Soviet Geography*, 8, No. 1 (1967): 12-24.

Semevskiy, B.N. "The Problems of Cuba's Energy Supplies." *Soviet Geography*, 8, No. 1 (1967): 25-33.

Semple, Hugh, Karyl Johnson, and Camilla Arjoonsingh. "Agrochemicals Usage and Protective Clothing Among Small Farmers in Grenada and Jamaica." *Caribbean Geography*, 14, No. 1 (2005): 15-30.

Shaw, Earl B. "The Fishing Industry of the Virgin Islands of the United States." *Bulletin of the Geographical Society of Philadelphia*, 31 (1933): 61-72.

Shaw, Earl B. "The Balanced Economy of St. John's Island." *Economic Geography*, 9, No. 2 (1933): 160-166.

Shaw, Earl B. "St. Croix: A Marginal Sugar Producing Island." *Geographical Review*, 23 (1933): 414-422.

Shaw, Earl B. "The Food Front in the Greater Antilles." *Economic Geography*, 19, No. 1 (1943): 55-76.

Shepard, C.Y. "Economic Survey of the Cacao Industry of Trinidad, British West Indies." *Economic Geography*, 3 (1927): 239-258.

Shepard, C.Y. "The Sugar Industry of the British West Indies and British Guiana with Special Reference to Trinidad." *Economic Geography*, 5 (1929): 149-175.

Skoczek, M. "The Role of Agricultural Cooperatives in Transformation of Rural Areas in Cuba in the 1980s." *Miscellanea Geographica* (Poland), 6 (1994): 227-234.

Sletto, Bjorn. "Fish, Rice, and the Meaning of Place: The Political Economy of the Nariva Swamp, Trinidad." *Caribbean Geography*, 9, No. 1 (2000): 14-29.

Smith, David A. "A Geographic Analysis of Inter-Island Transportation in the Lesser Antilles and Hawaiian Islands." Ph.D. Dissertation, University of Michigan, Ann Arbor, Mich., 1959.

Smith, Richard V. "Characteristics and Locational Problems of Puerto Rican Industrialization." Ph.D. Dissertation, Northwestern University, Evanston, Ill., 1957.

Smith, Samuel I. "A Federated British Caribbean: Resource Utilization." Master's Thesis, McGill University, Montreal, Quebec, 1955.

Smith, Samuel I. "Functional Ecology of Sugarcane in the American Tropics." *Caribbean Studies*, 15, No. 3 (1975): 57-77.

Smith, Samuel I. "Sugar Production Potential in Barbados." *Journal of Tropical Geography*, 42 (1976): 58-77.

Sofer, Michael, and Israel Drori. "Cooperation, Communalism, and Individualism: Rural Development and Land Settlement Schemes in Jamaica and Fiji." *Journal of Rural Cooperation*, 21, No. 2 (1993): 191-214.

Spaulding, Boyd. "Analysis of the Economic Dependency of the United States Virgin Islands on the Mother Country." Master's Thesis, Marshall University, Charleston, W.Va., 1964.

Spence, Balfour A.B. "Domestic-Food Production and Small Farming in Jamaica." Ph.D. Dissertation, University of Manitoba, Winnipeg, Manitoba, 1995.

Spence, Balfour A.B. "The Influence of Small Farmers' Land-Use Decisions on the Status of Domestic Food Security in Jamaica." *Caribbean Geography*, 7, No. 2 (1996): 132-142.

Spence, Balfour A.B. "Spatio-Evolutionary Model of Jamaican Small Farming." *Geographical Journal*, 165, No. 3 (1999): 296-305.

Spence, Balfour A.B., and Elizabeth Thomas-Hope. "Agrobiodiversity as an Environmental Management Tool in Small-Scale Farming Landscapes." Pp. 189-206 in *Caribbean Land and Development Revisited*, edited by J. Besson and Janet H. Momsen. New York, N.Y.: Palgrave Macmillan, 2007.

Sperberg, J. "Der Zuckersektor in Kuba zwischen Reform und Stillstand." Pp. 60-89 in *Kubas Weg aus der Krise: Neuorganisation der Produktion von Gutern und Dienstleistungen fur den Export*, edited by Gunter Mertins and Helmut Nuhn. Marburg, Germany: Marburger Geographische Schriften, Heft 138, 2001.

Stainfield, John. "The Fishing Industry of Jamaica." Master's Thesis, McGill University, Montreal, Quebec, 1974.

Starkey, Otis P. *The Economic Geography of Barbados: A Study of the Relationships Between Environmental Variations and Economic Development*. New York, N.Y.: Columbia University Press, 1939.

Starkey, Otis P. "The Economic Geography of Barbados." Ph.D. Dissertation, Columbia University, New York, N.Y., 1939.

Starkey, Otis P. "Declining Sugar Prices and Land Utilization in the British Lesser Antilles." *Economic Geography*, 18 (1942): 209-214.

Starkey, Otis P. *Commercial Geography of Montserrat*. Bloomington, Ind.: Indiana University, Department of Geography, Technical Report, No. 6, 1960.

Starkey, Otis P. *Commerical Geography of St. Kitts-Nevis*. Bloomington, Ind.: Indiana University, Department of Geography, Technical Report, No. 7, 1961.

Starkey, Otis P. *Commercial Geography of St. Lucia*. Bloomington, Ind.: Indiana University, Department of Geography, Technical Report, No. 8, 1961.

Starkey, Otis P. *Commercial Geography of Barbados*. Bloomington, Ind.: Indiana University, Department of Geography, Technical Report, No. 9, 1961.

Starkey, Otis P. *Commercial Geography of the Eastern British Caribbean*. Bloomington, Ind.: Indiana University, Department of Geography, Technical Report No. 12, 1961.

Starrett, Chris. "Export Processing Zones and Privatization: Development for the Caribbean." Master's Thesis, Ohio State University, Columbus, Ohio, 1998.

Steward, Julian H. "Perspectives on Plantations." *Revista Geografica*, 52 (1960): 77-85.

Stirling, Terri. "Alcan Jamaica Company and Small-Scale Agriculture: An Analysis of Competing Land Uses and Changing Land-Use Patterns." Master's Thesis, University of Regina, Regina, Saskatchewan, 1999.

Swanson, W.G. "A Linear Programming Analysis of Agricultural Diversification in Irrigable Areas of Barbados." Master's Thesis, Carleton University, Ottawa, Ontario, 1970.

Swisher, Carl L. "Agricultural Development Regions and Subregions of Puerto Rico." Ph.D. Dissertation, Northwestern University, Evanston, Ill., 1958.

Tata, Robert J. "Puerto Rican Economic Growth: Estimation of a Time Series Model in the Presence of Multicolinearity." *Journal of Tropical Geography*, 45 (1977): 58-64.

Tata, Robert J. "The Role of the Apparel Industry in Puerto Rico's Economic Development." *Southeastern Geographer*, 17, No. 2 (1977): 125-132.

Tata, Robert J. "General Impact Multiplier Analysis of Puerto Rico's Economy, 1965 to 1976." *Economic Geography*, 56, No. 1 (1980): 45-62.

Tata, Robert J. "The Relationship Between Puerto Rico's Economic Structure and Development." *Singapore Journal of Tropical Geography*, 2, No. 2 (1981): 129-135.

Tata, Robert J. "Combios en la Conducta Economica de Puerto Rico." *Revista Geografica*, 94 (1981): 101-115.

Terman, J. et al. "Land Capability Classification in Grenada, West Indies." *Mountain Research and Development*, 9, No. 1 (1989): 71-82.

Theisen, Kellen. "Assessing Land Degradation and Food Security in Haiti." Master's Thesis, University of Maryland, College Park, Md., 2002.

Thierbach, H. "Kubas wirtschaftsprobleme." *Zeitschaft fur Wirtschaftsgeographie*, 22, No. 4 (1978): 116-119.

Thomas-Hope, Elizabeth, and Adanna Jardine-Comrie. "Environmental Vulnerability and Ecosystem Services in the Jamaican Tourism Industry." Pp. 123-141 in *Global Change and Caribbean Vulnerability: Environment, Economy, and Society at Risk*, edited by Duncan McGregor, David Dodman, and David Barker. Mona, Jamaica: University of West Indies Press, 2009.

Thomas-Hope, Elizabeth, and Balfour A.B. Spence. "Jamaica." Pp. 270-292 in *Agrodiversity: Learning from Farmers Across the World*, edited by Harold Brookfield, Helen Parsons, and Muriel Brookfield. Tokyo, Japan: U.N. University Press, 2003.

Thomasson, David A. "Montserrat Kitchen Gardens: Social Functions and Development Potential." *Caribbean Geography*, 5, No. 1 (1994): 20-31.

Thomlinson, J.R. et al. "Land Use Dynamics in a Post-Agricultural Puerto Rican Landscape (1936-1988)." *Biotropic*,, 28 (1996): 525-536.

Thompson, Ronald A. "Small Scale Farming in St. Ann, Jamaica: A Cross Section Analysis of Gross Cash Income on a Selected Number of Farms." Master's Thesis, McMaster University, Hamilton, Ontario, 1970.

Timms, Benjamin. "Linkages Between Domestic Agriculture and the Hotel Sector in St. Lucia." Master's Thesis, Indiana University, Bloomington, Ind., 1999.

Timms, Benjamin. "Development Theory and Domestic Agriculture in the Caribbean: Recurring Crises and Missed Opportunities." *Caribbean Geography*, 15, No. 2 (2008): 101-117.

Toole, Steven T. "Land Use Change and Degradation in the Scotland District of Barbados." *Swansea Geographer*, 31 (1994): 106-115.

Torres, Rebecca M. "Introduction: Globalization, Development, and Environmental Change in the Caribbean." *Southeastern Geographer*, 45, No. 2 (2005): 165-173.

Torres, Rebecca M. et al. "Cuba's Farmers' Markets in the Special Period, 1990-1995." Pp. 53-68 in *Caribbean Land and Development Revisited*, edited by J. Besson and Janet H. Momsen. New York, N.Y.: Palgrave Macmillan, 2007.

Torres, Rebecca M. et al. "Experiment or Transition? Revisiting Food Distribution in Cuban Agromercadas from the Special Period." *Journal of Latin American Geography*, 9, No. 1 (2010): 67-87.

Tosi, Joseph. "Forest Land Utilization in Western Puerto Rico." Ph.D. Dissertation, Clark University, Worcester, Mass., 1959.

Tracey, Spencer. "Charity or Change? The Geography of NGOs in Jamaica's Development." Master's Thesis, Queen's University, Kingston, Ontario, 2001.

Traub-Werner, Marion. "Global Displacements: Geographies of Work and Industrial Restructuring in the Dominican Republic." Ph.D. Dissertation, University of Minnesota, Minneapolis, Minn., 2010.

Tuthill, R.L. "An Independent Farm in Cuba." *Economic Geography*, 25 (1949): 201-210.

Ungefehr, F. "Okonomische Abhangigkeit als Dilemma von Kleinstaaten: Das Beispiel der Tourismuswirtschaft der Bahamas." *Geographische Zeitschrift*, 76, No. 1 (1988): 48-60.

Uttley, Marguerite. "Land Utilization in the Canovanas Sugar District, Puerto Rico." Ph.D. Dissertation, University of Chicago, Chicago, Ill., 1937.

Vander Zang, Raymond. "We Do Not Yet Have Development: Encounters of Development Knowledges, Identities, and Practices in an NGO Program in Rural Haiti." Ph.D. Dissertation, Carleton University, Ottawa, Ontario, 2000.

Veenstra, A.W., H.M. Mulder, and R.A. Sels. "Analysing Container Flows in the Caribbean." *Journal of Transport Geography*, 13, No. 4 (2005): 295-305.

Venter, Jewell. "Possibilities of Rubber Production in Caribbean America." *Economic Geography*, 4, No. 4 (1928): 381-384.

Verge-Depre, R. "Quinze annees de conteneurisation des trafics maritimes aux Antilles francais." *Les Cahier d'Outre Mer*, 50 (1997): 151-170.

Vogt, J. "Quelques aspects de l'agriculture cubaine d'apres le Censo Agricola de 1946." *Information Geographique*, 18 (1954): 85-93.

Von Royen, William. "Transportation in the Dominican Republic." *Economic Geography*, 13 (1937): 287-299.

Wakefield, R.P. "Some Factors in Cuba's Foreign Trade." *Economic Geography*, 13 (1937): 109-131.

Wales, Robert W. "Landholdings and Agricultural Development in Puerto Rico: 1508-1970." Ph.D. Dissertation, University of Kansas, Lawrence, Kans., 1973.

Walker, Frederick. "Economic Progress of St. Vincent, British West Indies Since 1927." *Economic Geography*, 13 (1937): 217-234.

Warf, Barney. "Information Services in the Dominican Republic." *Yearbook, Conference of Latin Americanist Geographers*, 21 (1995): 13-24.

Weaver, David B. "Tourism as a Factor in Third World Development with Special Reference to the Caribbean." *Ontario Geography*, 22 (1983): 47-70.

Weber, G. "Der Transformationsprozess im Kubanischen Agrarsektor der 1990er Jahre." Pp. 191-222 in *Kubas Weg aus der Krise: Neu organisation der Produktion von Gutern und Dierstleistungen fur den Export*, edited by Gunter Mertins and Helmut Nuhn. Marburg, Germany: Marburger Geographische Schriften, Heft 138, 2001.

Weis, Anthony J. "On a Precipice: Globalization and Small Farmers in Eastern Jamaica." Ph.D. Dissertation, Queen's University, Kingston, Ontario, 2003.

Weiss, C.C. "Land Use in the Scotland District, Barbados." Master's Thesis, McGill University, Montreal, Quebec, 1966.

Welch, Barbara M. "Institutional Structures as a Factor in Land-Use Decisions: The Impact of Banana Growers' Associations in French and Commonwealth Islands in the Eastern Caribbean." Ph.D. Dissertation, University of London, London, England, 1989.

Welch, Barbara M. "Challenging Economic Irrelevance: The Role of Banana Growers' Associations in St. Lucia and Martinique." *Caribbean Geography*, 4, No. 2 (1993): 102-115.

Welch, Barbara M. *Survival by Association: Supply Management Landscapes of the Eastern Caribbean*. Montreal, Quebec: McGill-Queen's University Press, 1996. 364 pp.

Welch, Barbara M. "Table d'Hote or a la Carte? French and Commonwealth Approaches to Development in the Central Lesser Antilles." *Canadian Geographer*, 40, No. 2 (1996): 136-147.

Welch, Barbara M. "The Evolution of Export Banana Cultivation in Dominica and St. Lucia: Implications for Sustainability." *Caribbean Geography*, 7, No. 1 (1997): 62-74.

Werner, Marion. "Embodied Negotiations: Idenity, Space, and Livelihood After Trade Zones in the Dominican Republic." *Gender, Place, and Culture*, 17, No. 6 (2010): 725-741.

Wezel, A., and S. Bender. "Agricultural Land Use in the Coastal Area of the Alexander von Humbolt National Park, Cuba, and Its Implications for Conservation and Sustainability." *GeoJournal*, 57, No. 4 (2002): 241-249.

Whitbeck, R.H. "Geographical Relations in the Development of Cuban Agriculture." *Geographical Review*, 12 (1922): 223-241.

Whitbeck, R.H. "The Sugar Industry of Porto Rico." *Journal of Geography*, 29, No. 9 (1930): 363-370.

Whitbeck, R.H. "The Agricultural Geography of Jamaica." *Annals of the Association of American Geographers*, 22 (1932): 13-27.

White, R., G. Engelen, and I. Uljee. "Modeling Land Use Change with Linked Cellular Automata and Socio-Economic Models: A Tool for Exploring the Impact of Climate Change on the Island of St. Lucia." Pp. 189-204 in *Spatial Information for Land Use Management*, edited by M. Hill and Richard Aspinall. Amsterdam, Netherlands: Gordont Breach, 2000.

Whiteaker, Deanne. "Agricultural Land Use Change and Its Impact on Soil Resources in Southwest Antigua, West Indies, from 1964 to 1990." Master's Thesis, University of Oklahoma, Norman, Okla., 1994.

Wiley, James. "The Banana Industries of Costa Rica and Dominica in a Time of Change." *Tijdschrift voor Economische en Sociale Geographie*, 89, No. 1 (1998): 66-81.

Williams, Evon. "A Geographic Information System for the Management of Land and Bauxite Reserves, Study Area Kirkvine Manchester, Jamaica." Master's Thesis, University of Redlands, Redlands, Calif., 2003.

Williams, Ryan. "Influences of Spatial Variability of Gall Midge Incidence in Jamaican Hot Pepper Production." Master's Thesis, Virginia Polytechnic and State University, Blacksburg, Va., 2001.

Wills, Bernt L. "Land Utilization in the North Central Area of Puerto Rico: Bayamon to Vega Baja." Ph.D. Dissertation, Northwestern University, Evanston, Ill., 1953.

Wilson, Mark. "Jamaica's Back Office: Direct Dial Development or Dependency?" *Caribbean Geography*, 5, No. 2 (1994): 117-126.

Wilson, Mark. "The Natural Gas Industry of Trinidad and Tobago." *Caribbean Geography*, 7, No. 2 (1996): 143-150.

Winsberg, Morton D. "Isle of Pines: A Study in Economic Geography." Ph.D. Dissertation, University of Florida, Gainesville, Fla., 1958.

Winsberg, Morton D. "The Rise and Decline of the American Grapefruit Industry on the Isle of Pines, Cuba." *American Journal of Economics and Sociology*, 20 (1961): 543-548.

Wood, Harold A. "Departement du Nord, Haiti: A Study of Land Utilization and Settlement." Ph.D. Dissertation, University of Toronto, Toronto, Ontario, 1958.

Wood, Harold A. "Physical Influences on Peasant Agriculture in Northern Haiti." *Canadian Geographer*, 5, No. 2 (1961): 10-18.

Wood, Harold A. *Northern Haiti: Land, Land Use, and Settlement.* Toronto, Ontario: University of Toronto Press, 1963.

Wright, G. "Economic Conditions in St. Vincent, British West Indies." *Economic Geography*, 5 (1929): 231-259.

Wright, R.L. "Some Perspectives in Environmental Research for Agricultural Land-Use Planning in Developing Countries." *Geoforum*, 10 (1972): 15-34.

Yamaguchi, Takeshi. "Economic Development in the Former Territories in the West Indies Federation (in Japanese)." *Reports of the Geographical Institute, Komazawa University*, 4-5 (1968): 95-108.

Young, B.S. "Jamaica's Bauxite and Alumina Industries." *Annals of the Association of American Geographers*, 55, No. 3 (12965): 449-464.

Zornig, S. "Zucker macht abhangig: Erste Kuba ein Modell fur de Dritte Welt?" *Geographie Heute*, 7, No. 39 (1986): 41-45.

Zurbrigg, Elin. "Greening for Survival: The Economic Crisis, Urban Greening, and Sustainability in Havana, Cuba." *Middle States Geographer*, 32 (1999): 39-47.

Zweifer, Mark. "Land Use Evolution Along the Dominican Republic's Agricultural Frontier: A Comparative Analysis of Two Hill-Land Regions." Master's Thesis, Michigan State University, East Lansing, Mich., 1990.

Zweifer, Mark, Michael Gold, and Robert M. Thomas. *Land Use Evolution in Hill Regions of the Dominican Republic.* East Lansing, Mich.: Michigan State University, 1994.

Zweifer, Mark, Michael Gold, and Robert M. Thomas. "Land Use Evolution in Hill Regions of the Dominican Republic." *Professional Geographer*, 46, No. 1 (1994): 39-53.

PART 4

Environmental and Physical Geography

Acevedo-Gonzales, Miguel. "Geomorphologie und Hydrogeologie des Cuyaguateje-Flussgebiets, Pinar del Rio, Cuba." *Petermanns Geographische Mitteilungen*, 120 (1976): 177-191.

Acevedo-Gonzales, Miguel, and Carla Restrepo. "Land-Cover and Land-Use Change and Its Contribution to the Large-Scale Organization of Puerto Rico's Bird Assemblages." *Diversity and Distributions*, 14, No. 1 (2008): 114-122.

Acosta, Jessica. "Geographic Analysis of Habitat Change for Endemic Animal Species of Puerto Rico." Master's Thesis, Pennsylvania State University, University Park, Pa., 2007.

Ahmad, Rafi. "Primer on Earthquake Hazards and Disasters in Jamaica." *Caribbean Geographer*, 10, No. 2 (1999): 124-136.

Ahmad, Rafi. "Natural Hazard Maps in Jamaica." *Caribbean Geography*, 12, No. 2 (2001): 90-107.

Alexander, Charles S. "The Marine Terraces of Aruba, Bonaire, and Curacao, Netherlands Antilles." *Annals of the Association of American Geographers*, 51, No. 1 (1961): 102-123.

Alexander, W.H. "The Rainfall of St. Kitts." *Monthly Weather Review*, 27 (1899): 583-587.

Alexander, W.H. "The Flora of St. Christopher." *Bulletin of the American Geographical Society*, 33, No. 3 (1901): 207-219.

Alexander, W.H. "The Climatology of Antigua." *Monthly Weather Review*, 29 (1901): 165-167.

Alexander, W.H. "Porto Rico: Its Climate and Resources." *Bulletin of the American Geographical Society*, 34, No. 5 (1902): 401-409.

Andereck, Zachary. "Mapping Vulnerability of Infrastructure to Destruction by Slope Failure on the Island of Dominica, West Indies: A Case Study of Good Fond, Petite Soutriere, and Moune Jaune." Master's Thesis, Miami University, Oxford, Ohio, 2007.

Anderson, Malcolm. "Road-Cut Slope Topography and Stability Relationships in St. Lucia, West Indies." *Applied Geography*, 3, No. 2 (1983): 105-114.

Anderson, Malcolm, and P.E. Kneale. "An Examination of the Relationships Between Storm Precipitation and Poor Water Conditions in Road Cut Slopes, St. Lucia, West Indies." *Singapore Journal of Tropical Geography*, 1, No. 1 (1980): 1-8.

Anderson, T. "Recent Volcanic Eruptions in the West Indies." *Geographical Journal*, 21 (1903): 265-281.

Andre, E. "The Volcanic Eruption at St. Vincent." *Geographical Journal*, 20 (1902): 6068.

Andrews, Alexa J. "Spatial and Temporal Variability of Tropical Storm and Hurricane Strikes in the Bahamas and the Greater and Lesser Antilles." Master's Thesis, Louisiana State University, Baton Rouge, La., 2007.

Antonini, Gustavo. "Process and Patterns of Landscape Change in the Linea Noreste, Dominican Republic." Ph.D. Dissertation, Columbia University, New York, N.Y., 1968.

Arroyo-Tuttup, Maria R. "Soil Erosion and Conservation Measures in Cibuco District, Puerto Rico." Master's Thesis, University of Texas, Austin, Tex., 1989.

Ayres, Christopher. "Landscape Fragmentation Analysis on the Island of Hispanola from 1986-2003." Master's Thesis, Clark University, Worcester, Mass., 2005.

Baban, Serwan, Anthony Ramdath, and J. Agard. "Examining Climate Change in Small Tropical Islands Using Remotely-Sensed Data: A Case Study of Trinidad and Tobago." *Caribbean Geography*, 14, No. 2 (2005): 91-101.

Balcerzak, William. "Selected Water Quality Parameters of a Carbonate Island Karst Aquifer, San Salvador Island, Bahamas." Master's Thesis, Mississippi State University, State College, Miss., 1991.

Barker, David. "Land Degradation in the Yallahs Basin, Jamaica: Historical Notes and Contemporary Observations." *Geography*, 73 (1988): 116-124.

Barker, David. "The North-East Trades and Temperature Inversion: Notes for the Classroom." *Caribbean Geography*, 9, No. 2 (1998): 58-65.

Barrett, W. "Emerged and Submerged Shorelines of the Dominican Republic." *Revista Geografica*, 30 (1962): 51-77.

Basnayake, Ballage K. "Two Maps of Direct Short-Wave Radiation in Barbados." *Climatological Bulletin*, 4 (1968): 21-30.

Basnayake, Ballage K. "Topography and Solar Radiation in Barbados." Ph.D. Dissertation, McGill University, Montreal, Quebec, 1971.

Beach, Timothy et al. "A Soils History of Mesoamerica and the Caribbean Islands." Pp. 51-90 in *Soils and Societies: Perspectives from Environmental History*, edited by J.R. McNeill and V. Winiwarter. Isle of Harris, Scotland: White Horse Press, 2006.

Beard, John S. *The Natural Vegetation of the Windward and Leeward Islands.* Oxford, England: Oxford Forestry Memoirs, No. 21, 1949.

Beaulieu, A. "The Changing Plant Association of Martinique from Pre-Columbian Times to the Present Day." Master's Thesis, McGill University, Montreal, Quebec, 1965.

Beckford, Yvette H.M. "The Effects of Storm Surges on Caribbean Coastal Areas: The Case of Hurricane Lenny." *Caribbean Geography*, 11, No. 2 (2000): 100-107.

Bellamy, C.V. "The Rainfall of Dominica." *Quarterly Journal of the Royal Meteorological Society*, 23 (1897): 261-275.

Bellamy, C.V. "The Rainfall of Dominica." *Quarterly Journal of the Royal Meteorological Society*, 29 (1903): 23-28.

Bennett, H.H. "Some Geographical Aspects of Cuban Soils." *Geographical Review*, 18, No. 1 (1928): 68-82.

Besoncenot, J. "Introduction a l'etude geographique du bioclimat humain des petites Antilles francaises." *Bulletin de l'Association de Geographes Francais*, 47 (1970): 167-179.

Blanchon, Paul. "Controls on Modern Reef Development Around Grand Cayman." Ph.D. Dissertation, University of Alberta, Edmonton, Alberta, 1995.

Blanchon, Paul, and B. Jones. "Marine-Planation Terraces on the Shelf Around Grand Cayman: A Result of Stepped Holocene Sea-Level Rise." *Journal of Coastal Research*, 11 (1995): 1-33.

Blaunt, James M. "A Study of Cultural Determinants of Soil Erosion and Conservation in the Blue Mountains of Jamaica: Progress Report." *Social and Economic Studies*, 8 (1959): 403-420.

Blick, J. "Pre-Columbian Impact on Terrestrial, Intertidal, and Marine Resources, San Salvador, Bahamas (AD 950-1500)." *Journal for Nature Conservation*, 15, No. 3 (2007): 174-183.

Boruff, Jeffrey. "A Multiple Hazards Assessment of Two Caribbean Nations: Barbados and St. Vincent." Ph.D. Dissertation, University of South Carolina, Columbia, S.C., 2005.

Boruff, Jeffrey, and Susan L. Cutter. "The Environmental Vulnerability of Caribbean Island Nations." *Geographical Review*, 97, No. 1 (2007): 24-45.

Bowden, M.J. *Climate, Water Balance, and Climatic Change in the North-West Virgin Islands*. St. Thomas, Virgin Islands: Caribbean Research Institute, 1970.

Bowden, M.J. *Hurricane in Paradise: Perception and Reality of the Hurricane Hazard in the Virgin Islands*. St. Thomas, Virgin Islands: Island Resources Foundation, 1974.

Bowden, M.J. et al. *Water Balance of a Dry Island: The Hydroclimatology of St. Croix, Virgin Islands, and Potential for Agriculture and Urban Growth*. Worcester, Mass.: Clark University, Geography Department, 1968.

Bowman, Robert G. "Soil Erosion in Puerto Rico." Ph.D. Dissertation, University of California, Berkeley, Calif., 1941.

Bravaro, J. et al. "Les effets des glissements de terrain sur la morphodynamique fluviale dans le basin de la Layou (La Dominique, Antilles)." *Geomorphologie*, 4 (2001): 257-270.

Brock, A., and B. Buck. "Polygenetic Development of the Mormon Mesa, NV Petrocalcic Horizons: Geomorphic and Paleoenvironment of Interpretations." *Catena*, 77, No. 1 (2009): 65-76.

Brockmann, Vernon W. "Physical Land Types and Land Utilization in the Caguas-San Lorenzo Region of Puerto Rico." Ph.D. Dissertation, Northwestern University, Evanston, Ill., 1951.

Brooks, C.E.P. "The Meteorology of Nassau, Bahamas, 1852-1919." *Quarterly Journal of the Royal Meteorological Society*, 47 (1921): 59-62.

Brooks, S.M., and Fiona F. Whitaker. "Geochemical and Physical Controls on Vadose Zone Hydrology of Holocene Carbonate Sands, Grand Bahama Island." *Earth Surface Processes and Landforms*, 22, No. 1 (1997): 45-58.

Brown, Dwight A. "Erosion and Morphometry of Small Drainage Basins in Eastern Puerto Rico." Ph.D. Dissertation, University of Kansas, Lawrence, Kans., 1968.

Byrne, A. Roger. "Man and the Variable Vulnerability of Island Life: A Study of Recent Vegetation Change in the Bahamas." Ph.D. Dissertation, University of Wisconsin, Madison, Wis., 1972.

Byron, Margaret. "Collateral and Achievement: Land and Caribbean Migration." Pp. 243-254 in *Caribbean Land and Development Revisited*, edited by J. Besson and Janet H. Momsen. New York, N.Y.: Palgrave Macmillan, 2007.

Camara, R. "Les savanes de la Republique Dominicaine et Leur exploitation par les system 'hatos.'" *Les Cahiers d'Outre Mer*, 53, No. 212 (2000): 343-366.

Carlson, F.A. "Wind Directions: Puerto Rico." *Journal of Geography*, 47 (1948): 372-374.

Carson, M.A., and S.W. Tom. "The Land Conservation Conundrum of Eastern Barbados." *Annals of the Association of American Geographers*, 67, No. 2 (1977): 185-203.

Carter, M.M., J.B. Elsner, and S.P. Bennett. "A Quantitative Precipitation Forecast Experiment for Puerto Rico." *Journal of Hydrology*, 239, No. 1-4 (2000): 162-178.

Caviedes, Cesar N. "Five Hundred Years of Hurricanes in the Caribbean: Their Relationship with Global Climatic Variabilities." *GeoJournal*, 23, No. 4 (1991): 301-310.

Chapman, V.J. "The Botany of the Jamaica Shoreline." *Geographical Journal*, 96 (1940): 312-324.

Chiappy-Jhones, C. et al. "Floristic Affinities Between the Yucatan Peninsula and Some Karstic Areas of Cuba." *Journal of Biogeography*, 28, No. 4 (2001): 535-542.

Chin, L.S. "Albedo Measurements of Various Surfaces in Barbados." Master's Thesis, McGill University, Montreal, Quebec, 1964.

Chmura, Gail L., Amanda Coffey, and Ryan Crago. "Effects of Sea Spray on Vadose Diagenesis of Late Quaternary Aeolianites, Bermuda." *Journal of Coastal Research*, 17, No. 1 (2001): 228-240.

Christian, C.S. "The Challenge of Parrot Conservation in St. Vincent and the Grenadines." *Journal of Biogeography*, 20, No. 5 (1993): 463-470.

Christian, C.S. et al. "Parrot Conservation and Ecotourism in the Windward Islands." *Journal of Biogeography*, 23, No. 3 (1996): 387-394.

Clark, G.M., S.P. Horn, and Kenneth H. Orvis. "High-Elevation Savanna Landscapes in the Cordillera Central, Dominican Republic, Hispanola: Dambos in the Caribbean?" *Mountain Research and Development*, 22 (2002): 288-295.

Collymore, J.M., and M.D. Griffith. "Flooding in Speightstown: Towards a Flood Management Strategy." Pp. 117-125 in *Proceedings of a Meeting of Experts on Hazard Mapping in the Caribbean*, edited by David Barker. Mona, Jamaica: University of the West Indies, Department of Geography, 1987.

Cook, Benjamin et al. "Immigration History of Amphidromous Species on a Greater Antillian Island." *Journal of Biogeography*, 37, No. 2 (2010): 270-277.

Copland, K.K. "Local Ecological Knowledge of Coral Reef Ecosystems: A Case Study of Bequia, St. Vincent, and the Grenadines." Master's Thesis, Carleton University, Ottawa, Ontario, 2009.

Cornish, V. "The Jamaican Earthquake (1907)." *Geographical Journal*, 31, No. 3 (1908): 245-271.

Cotilla Rodriquez, Merio O. "Estudio geomorfologico del carso en la region Gibara-Banes de Holguin, Cuba." *Revista Geografica*, 134 (2003): 9-28.

Cotilla Rodriquez, Merio O., and Diego C. Barba. "Determinants of Linements in Hispanola." *Revista Geografica*, 147 (2010): 133-154.

Cotilla Rodriquez, Merio O. et al. "La red fluvial de Cuba y su interpretacion morfoestructural." *Revista Geografica*, 134 (2003): 29-46.

Cross, John A. "Natural Hazards within the West Indies." *Journal of Geography*, 91, No. 5 (1992): 190-199.

Crump, Michael A., and Douglas W. Gamble. "Hydroclimatic Analysis of a Carbonate Island Pond Through the Development of a Hydrologic Landscape Unit Model." *Physical Geography*, 27, No. 6 (2006): 554-570.

Cucurullo, Oscar. "Rasgas sobre la orogenesis y la topografia de la Republica Dominicana." *Revista Geografica*, 9-10 (1950): 125-132.

Darsan, Junior. "A Comparative Study of the Coastal Morphology of Cocos Bay and Las Vuevas Bay, Trinidad." *Caribbean Geography*, 14, No. 2 (2005): 116-132.

Davidson, Dawn E. "The Modern Pollen Spectra of Mangrove Dominated Plant Communities in the Sabana-Camaguey Archipelago, Cuba." Master's Thesis, University of Toronto, Toronto, Ontario, 2008.

Davis, Lisa et al. "Observations of Environmental Change in Cuba." *Southeastern Geographer*, 49, No. 1 (2009): 84-93.

Davolos, Liliana M. "Short-Faced Bats (Phyllostomidae: Stenodermatina): A Caribbean Radiation of Strict Frugivores." *Journal of Biogeography*, 34, No. 2 (2007): 364-275.

Day, Michael J. "The Morphology and Hydrology of Some Jamaican Karst Depressions." *Earth Surface Processes*, 1 (1976): 111-130.

Day, Michael J. "The Hydrology of Polygonal Depressions in Northern Jamaica." *Zeitschrift fur Geomorphologie*, Supplement, 32 (1979): 25-34.

Day, Michael J. "Doline Morphology and Development in Barbados." *Annals of the Association of American Geographers*, 73, No. 2 (1983): 206-219.

Day, Michael J. "Predicting the Location of Surface Collapse Within Karst Depressions: A Jamaican Example." Pp. 137-151 in *Sinkholes: Their Geology, Engineering and Environmental Impact*, edited by B.F. Beck. Rotterdam, Netherlands: Balkema, 1984.

Day, Michael J. "An Assessment to Karstic Collapse Hazards at Mount Rosser, Ewarton, Jamaica." Pp. 40-49 in *Sinkholes and the Engineer-*

ing, and Environmental Impacts of Karst, edited by B.F. Beck. American Society of Civil Engineers, Geotechnical Special Publication, No. 122, 2003.

Day, Michael J. "Natural and Anthropogenic Hazards in the Karst of Jamaica." Pp. 173-184 in *Natural and Anthropogenic Hazards in Karst Areas: Recognition, Analyses, and Mitigation*, edited by M. Parise and J. Gunn. London, England: The Geological Society, Special Publication, No. 279, 2007.

Day, Michael J. *Karst Landscapes of the Caribbean: Their Use and Protection*. Coconut Creek, Fla.: Caribbean Studies Press, 2009.

Day, Michael J., and M.S. Chenoweth. "The Karstlands of Trinidad and Tobago, Their Land Use, and Conservation." *Geographical Journal*, 170, No. 3 (2004): 256-266.

Day, Michael J., and M.S. Chenoweth. "Potential Impacts of Anthropogenic Environmental Change on the Caribbean Karst." Pp. 100-122 in *Global Change and Caribbean Vulnerability: Environment, Economy, and Society at Risk*, edited by Duncan McGregor, David Dodman, and David Barker. Mona, Jamaica: University of West Indies Press, 2009.

Day, Michael J., and S. Koenig. "Cave Monitoring Priorities in Central America and the Caribbean." *Acta Carsologica*, 30, No. 1 (2002): 123-134.

De Graaf, Jan. "Determining the Significance of Landslide Activity: Examples from the Eastern Caribbean." *Caribbean Geography*, 3 (1991): 29-42.

Delahuntz, J.L. "Remote Sensing as a Tool for Mapping and Monitoring Habitat: A Case Study of the St. Vincent Parrot." *Caribbean Geography*, 14, No. 1 (2005): 31-39.

Desjardins, Amos. "Paleoenvironmental Reconstruction of a Coastal Lagoon in Southwestern Dominican Republic." Master's Thesis, Virginia Polytechnic and State University, Blacksburg, Va., 2007.

Dexter, K.G. "The Influence of Dispersal on Macroecological Patterns of Lesser Antillean Birds." *Journal of Biogeography*, 37, No. 11 (2010): 2137-2147.

Doerr, Arthur H., and Don R. Hoy. "Karst Landscapes of Cuba, Puerto Rico, and Jamaica." *Scientific Monthly*, 85 (1957): 178-187.

Doran, Edwin, Jr. "Land Forms of Grand Cayman Island, British West Indies." *Texas Journal of Science*, 6 (1954): 360-377.

Doran, Edwin, Jr. *Landforms of the Southern Bahamas*. Austin, Tex.: University of Texas Press, 1955.

Dragoni, Alexis L. "GIS Assisted Direct Gradient Analysis for Predicting High Elevation Forest Associations in the Taro Negro Region of Puerto Rico." Master's Thesis, Ohio University, Athens, Ohio, 1992.

Dunfield, Peter. "Some Effects of a Sugar-Factory Byproduct Compost on Root Growth and Mycorrhisal Infection of Sugarcane in Barbados." Master's Thesis, McGill University, Montreal, Quebec, 1991.

Durland, W.D. "The Forests of the Dominican Republic." *Geographical Review*, 12 (1922): 206-222.

El Fouladi, Abderrahman. "Hausse de niveau moyen relatif de la mer a Trinidad (Caraibes): Evidence, causes probables et evaluation." Ph.D. Dissertation, University of Montreal, Montreal, Quebec, 2006.

Engel, M. et al. "Reading the Chapter of Extreme Wave Events in Nearshore Geo-Bio-Archives of Bonaire (Netherlands Antilles): Initial Results from Lagua and Boka Bartol." Pp. 157-178 in *Ergebnisse aktueller Kuntenforschung. Beitrage der 26. Jahrestagung des Arbeitskreis Geographie der Mere und Kusten, 25-27 April 2008*, edited by A. Votl and H. Bruckner. Marburg, Germany: Marburger Geographische Schriften, Heft 145, 2009.

Ewaschuk, Carolyn. "Soil Survey and Land Suitability Classification with Application of Geographic Information Systems: Union Island, St. Vincent, and the Grenadines." Master's Thesis, University of California, Los Angeles, Calif., 1995.

Eyre, L. Alan. "Vegetation Change and Desertification in the Caribbean." *Proceedings of International Satellite Land Surface Climatology Project (ISLSCP) Conference*, Rome, December, 1985. European Science Association, Special Publication No. 240 (1986): 509-514.

Eyre, L. Alan. "Jamaica: Test Case for Tropical Deforestation?" *Ambio*, 16 (1987): 336-341.

Eyre, L. Alan. "Deforestation in Jamaica." *Journal of the Scientific Research Council of Jamaica*, 6 (1987): 17-24.

Eyre, L. Alan. "Slow Death of a Tropical Rain Forest: The Cockpit Country of Jamaica, West Indies." Pp. 599-606 in *Environmental*

Quality and Ecosystem Stability, edited by M. Lurin, Y. Steinberger, and E. Spanier. Jerusalem, Israel: Israel Society for Ecology and Environmental Quality Publication, 1989.

Eyre, L. Alan. "Hurricane Gilbert: Caribbean Recent Breaker." *Weather*, 48 (1989): 160-169.

Eyre, L. Alan. *Depletion of the Jamaican Rain Forest: The Social Cultural Consequences*. Quito, Ecuador: Universidad San Francisco de Quito and the Association of Academic Progress in Latin America and the Caribbean, 1992.

Eyre, L. Alan. "The Effects of Environmental Degradation in the Cane River and Rio Minho Watersheds, Jamaica: A Commentary." Pp. 57-65 in *Natural Hazards in the Caribbean*, edited by Rafi Ahmad. Jamaica: *Journal of the Geological Society of Jamaica*, Special Issues No. 12, 1992.

Eyre, L. Alan. "The Tropical Rainforests of the Eastern Caribbean: Present Status and Conservation." *Caribbean Geography*, 9, No. 2 (2000): 101-120.

Fassig, D.L. "The Climate of Porto Rico." *Annals of the Association of American Geographers*, 1 (1911): 127-134.

Fermor, John. *Northers and Weather at Kingston, Jamaica*. Mona, Jamaica: University of the West Indies, Department of Geography, Paper No. 2, 1970.

Fermor, John. *The Dry Valleys of Barbados: Their Pattern and Origin, A Critical View*. Mona, Jamaica: University of the West Indies, Department of Geography, Paper No. 6, 1971.

Fermor, John. "The Weather During Northers at Kingston, Jamaica." *Journal of Tropical Geography*, 31 (1971): 31-37.

Fermor, John. "The Dry Valleys of Barbados: A Critical Review of Their Pattern and Origin." *Transactions, Institute of British Geographers*, 57 (1972): 153-166.

Fleurant, C. et al. "A Model of Cockpit Karst Landscape, Jamaica." *Geomorphologie*, 1 (2008): 3-14.

Fonaroff, L. Schuyler. "Urbanization, Birds, and Ecological Change in Northeastern Trinidad." *Biological Conservation*, 6, No. 4 (1974): 258-262.

Frank, E. "Aspects of Karst Development and Speleogenesis on Isla de Mona, Puerto Rico: An Analogue for Pleistocene Speleogenesis in the Bahamas." Master's Thesis, Mississippi State University, State College, Miss., 1993.

Frost, David B. "Drag Coefficients Over Agricultural Surfaces in Montserrat, West Indies." *Proceedings, New England-St. Lawrence Valley Division, Association of American Geographers*, 16 (1986): 63-72.

Furland, W.D. "Forest Regeneration in Puerto Rico." *Economic Geography*, 5 (1929): 369-381.

Gamble, Douglas W. "Caribbean Vulnerability: Development of an Appropriate Climatic Framework." Pp. 22-48 in *Global Change and Caribbean Vulnerability: Environment, Economy, and Society at Risk*, edited by Duncan McGregor, David Dodman, and David Barker. Mona, Jamaica: University of West Indies Press, 2009.

Gamble, Douglas W., and Scott Curtis. "Caribbean Precipitation: Review, Model, and Prospect." *Progress in Physical Geography*, 32, No. 3 (2008): 265-276.

Gamble, Douglas W. et al. "Spatial Variability of the Caribbean Mid-Summer Drought and Relation to the North Atlantic High." *International Journal of Climatology*, 28, No. 3 (2008): 343-350.

Gellis, Allen C. et al. "Land-Use Effects on Erosion, Sediment Yields, and Reservoir Sedimentation: A Case Study in the Lago Loiza Basin, Puerto Rico." *Physical Geography*, 27, No. 1 (2006): 39-69.

Gillis, W.T. "Biogeography and Vegetation." Pp. 13-23 in *Land Resources of the Bahamas: A Summary*, edited by B.G. Little et al. England: Ministry of Overseas Development, Land Resources Study, No. 27, 1977.

Gotelli, N.J., and L.G. Abele. "Statistical Distributions of West Indian Bird Families." *Journal of Biogeography*, 9, No. 5 (1982): 421-435.

Gould, Joseph, and Stephen Vermette. "Characterizing the Beach Morphology of San Salvador, Bahamas." *Middle States Geographer*, 38 (2005): 61-68.

Granger, Orme E. "Climatic Fluctuations in Trinidad, West Indies, 1921-1966." Master's Thesis, University of Alberta, Edmonton, Alberta, 1971.

Granger, Orme E. "The Hydroclimatonomy of a Developing Tropical Island: A Water Resource Perspective." *Annals of the Association of American Geographers*, 73, No. 2 (1983): 183-205.

Granger, Orme E. "Caribbean Climates." *Progress in Physical Geography*, 9, No. 1 (1985): 16-43.

Granger, Orme E. "Caribbean Island States: Perils and Prospects in a Changing Global Environment." *Journal of Coastal Research*, 24 (1997): 71-93.

Green, E.P. et al. "Remote Sensing Techniques for Mangrove Mapping." *International Journal of Remote Sensing*, 19, No. 5 (1998): 935-956.

Green, E.P. et al. "The Assessment of Mangrove Areas Using High Resolution Multispectral Airborne Imagery." *Journal of Coastal Research*, 14, No. 2 (1998): 433-443.

Grossman, Lawrence S. "Pesticides, People, and the Environment in St. Vincent." *Caribbean Geography*, 3, No. 3 (1992): 175-186.

Grossman, Lawrence S. "Pesticides, Caution, and Experimentation in St. Vincent, Eastern Caribbean." *Human Ecology*, 20, No. 3 (1993): 315-336.

Guernsey, Jackson E. "The Climate of Jamaica." Ph.D. Dissertation, Clark University, Worcester, Mass., 1941.

Gupta, Avijit. "Effect of Seasonal Glow and High Magnitude Floods on Channel Form and Stream Behavior in Eastern Jamaica." Ph.D. Dissertation, Johns Hopkins University, Baltimore, Md., 1973.

Hagey, Frances. "Analysis of Pleistocene Lake and Lagoon Deposits, San Salvador Island, Bahamas." Master's Thesis, Mississippi State University, State College, Miss., 1991.

Halkitis, M., Samuel I. Smith, and K. Rigg. *The Climate of the Bahamas*. Nassau, Bahamas: Bahamas Geographical Association, 1980.

Hall, Robert B. "The Island of Gonave: A Study in Karst Landscape." *Papers, Michigan Academy of Science*, 10 (1928): 161-168.

Hall, Robert B., and H.M. Kendall. "The Climates of the Republic of Haiti." *Papers, Michigan Academy of Science, Arts, and Letters*, 14 (1931): 351-366.

Harden, Carol P., and Paul Delmas Scruggs. "Infiltration on Mountain Slopes: A Comparison of Three Environments." *Geomorphology*, 55, No. 1-4 (2003): 5-24.

Harris, David R. "Plants, Animals, and Man in the Outer Leeward Islands, West Indies: An Ecological Study of Antigua, Barbuda, and Anguilla." Ph.D. Dissertation, University of California, Berkeley, Calif., 1963.

Harris, David R. *Plants, Animals and Man in the Outer Leeward Islands, West Indies*. Berkeley, Calif.: University of California, Publications in Geography, Volume 18, 1965. 163 pp.

Heatwole, H., and R. Levin. "Biogeography of the Puerto Rican Bank: Species Turnover on a Small Cay, Cayo Ahogedo." *Ecology*, 54 (1973): 1042-1055.

Helmer, E.H. et al. "Mapping the Forest Type and Land Cover of Puerto Rico: A Component of Caribbean Biodiversity Hotspot." *Caribbean Journal of Science*, 38, No. 3-4 (2002): 165-183.

Hendry, Malcolm D. "The Influence of the Sea-Land Breeze Regime on Beach Erosion and Accretion: An Example from Jamaica." *Caribbean Geography*, 1, No. 1 (1983): 13-24.

Hernandez-Leal, P.A. et al. "Analysis of Vegetation Patterns in the Hispanola Island Using Advanced Very High Resolution Radiometer (AVHRR) Data." *Advances in Space Research*, 38 (2006): 2203-2207

Hodge, W.H. "The Vegetation of Dominica." *Geographical Review*, 33 (1943): 349-375.

Horst, Oscar H. "Climate and the Encounter in the Dominican Republic." *Journal of Geography*, 91, No. 5 (1992): 205-210.

Hovey, Edmund O. "Ten Days in Camp on Mt. Pele, Martinique: The Volcano Six Years After the Great Eruption." *Bulletin of the American Geographical Society*, 40, No. 11 (1908): 662-679.

Howard, Richard A. *The Vegetation of the Grenadines, Windward Islands, British West Indies*. Cambridge, England: Gray Herbarium of Harvard University, 1952.

Howard, Richard A., and George R. Procter. "Vegetation in Bauxite Soils in Jamaica." *Journal of the Arnold Arboretum*, 38 (1957): 1-41, and 151-169.

Hubbard, Raymond, and John Fermor. "Land-Slides on Jamaican Roads: An Appraisal of Causes." *Geographical Survey*, 1, No. 3 (1972): 16-30.

Hubbard, Raymond, and John Fermor. *Landslides on Jamaican Roads: An Appraisal of Causes*. Mona: University of the West Indies, Department of Geography, Paper No. 7, 1972.

Huetz de Lemps, C. "L'homme et le vegetation a la Barbade." *Les Caheirs d'Outre Mer*, 20 (1967): 406-409.

Izzo, Michela et al. "A New Climatic Map of the Dominican Republic Based on the Thornthwaite Classification." *Physical Geography*, 31, No. 5 (2010): 455-472.

James, Arlington, and Ted Serrant. "The Carholm-Huxley Landslides: An Assessment of Their Impacts on the Lower Layou River Valley, Dominica." *Caribbean Geography*, 8, No. 2 (1997): 120-127.

James, Preston E. "The Climate of Trinidad, B.W.I." *Monthly Weather Review*, 53 (1925): 71-75.

Jessen, Catherine et al. "A Late Holocene Palaeoenvironmental Record from Altona Bay, St. Croix, U.S. Virgin Islands." *Geografisk Tidsskrift*, 108, No. 2 (2008): 50-70.

Jimenez, A. et al. "Der Einfluss geomorphologischer Prozesse auf die Umlagerung von Sediments-, Verttierungs-, und Bodendecken in den west kubanischen Kustenebenen." *Petermanns Geographische Mitteilungen*, 121, No. 2 (1977): 111-122.

Kalbfleish, William. "Sedimentology of Frank Sound and Pease Bay, Two Modern Shallow-Water Hurricane-Affected Lagoons, Grand Cayman, British West Indies." Ph.D. Dissertation, University of Alberta, Edmonton, Alberta, 1995.

Kalbfleish, William, and B. Jones. "Sedimentology of Shallow Hurricane-Affected Lagoons: Grand Cayman, British West Indies." *Journal of Coastal Research*, 14 (1998): 140-161.

Kan, H.T.P. "Soil Conservation and Land Use in Scotland District of Barbados." Master's Thesis, McGill University, Montreal, Quebec, 1964.

Kappas, M. "Klimaokologische Aspekte eines Bergwaldgebiets in der Dominikanischen Republik." *Geographische Rundschau*, 51, No. 9 (1999): 462-468.

Kelletat, D. et al. "Holocene Tsunami Deposits on the Bahaman Islands of Long Island and Eleuthera." *Zeitschrift fur Geomorphologie*, 48, No. 4 (2004): 519-540.

Kennedy, Laura L. "The Potential of Agro-Forestry Systems in Controlling Soil Erosion on Sloping Lands, St. Vincent, West Indies." Master's Thesis, University of Calgary, Calgary, Alberta, 1999.

Kennedy, Lisa M. "Fire and Forest in the Highlands of the Cordillera Central, Dominican Republic: Modern Dynamics and Long-Term History." Ph.D. Dissertation, University of Tennessee, Knoxville, Tenn., 2003.

Keshishev, V.N. "The Runoff in Eastern Cuba." *Soviet Geography*, 8, No. 1 (1967): 1-11.

Kimber, Clarissa. "Recent Historical Plant Geography of Martinique." Ph.D. Dissertation, University of Wisconsin, Madison, Wis., 1968.

Knowles, Jason T. "Coastal Lake-Sediment Records of Prehistoric Hurricane Strikes in Honduras and Turks and Caicos Island of the Caribbean Basin." Ph.D. Dissertation, Louisiana State University, Baton Rouge, La., 2008.

Kobara, Shinichi. "Regional Analysis of Seafloor Characteristics at Reef Fish Spawning Aggregation Sites in the Caribbean." Ph.D. Dissertation, Texas A&M University, College Station, Tex., 2009.

Kohsick, L., C.H. Hulsberger, and J.H.J. Terwindt. "Beach Erosion Along the West Coast of Aruba, Netherlands Antilles." *Journal of Coastal Research*, 3, No. 1 (1987): 37-53.

Kramer, Daniel E. "Holocene Carbonate Deposits in a Transitional Terrestrial to Marine Setting, Blackwood Bay, San Salvador Island, Bahamas." Master's Thesis, Mississippi State University, State College, Miss., 1990.

Kueny, Jeffrey A. "An Assessment of Protected Karst Landscapes in the Caribbean." *Caribbean Geography*, 9, No. 2 (1998): 87-100.

Lackhai, Hayden. "A Comparative Study of the Coastal Geomorphology of San Souci, Saline, Quinan, and Carenage Bays in Trinidad." *Caribbean Geography*, 10, No. 2 (1999): 103-123.

Lackhai, Hayden. "Coastal Energy Inputs, Beach Morphology, and Sedimentological Characteristics in Trinidad." *Caribbean Geography*, 11, No. 1 (2000): 34-43.

Lahey, James F. "On the Origin of the Dry Climate in Northern South America and the Southern Caribbean." Ph.D. Dissertation, University of Wisconsin, Madison, Wis., 1958.

Landis, Kate. "Assessing Tropical Deforestation Rates Using Landsat Multispectral Scanner Data: A Case Study of Haiti." Master's Thesis, University of Georgia, Athens, Ga., 1994.

Landmann, M. *Reliefgerationen und Formengenese im Gebiet des Lluidas Vale-Poljes/Jamaika.* Tubingen, Germany: Tubinger Geographische Studien, Volume 101, 1989.

Larsen, Matthew C. "Tropical Geomorphology and Geomorphic Work: Study of Geomorphic Processes and Sediment and Water Budgets in Montane Humid-Tropical Forested and Developed Watersheds, Puerto Rico." Ph.D. Dissertation, University of Colorado, Boulder, Colo., 1997.

Larsen, Matthew C. "Analysis of Twentieth Century Rainfall and Streamflow to Characterize Drought and Water Resources in Puerto Rico." *Physical Geography*, 21, No. 6 (2000): 494-521.

Larsen, Matthew C., and A. Simon. "A Rainfall Intensity-Duration Threshold for Landslides in a Humid-Tropical Environment, Puerto Rico." *Geografiska Annaler A*, 75A, No. 1-2 (1993): 13-24.

Lasserre, Guy. "Notes sur le karst de la Guadeloupe." *Erdkunde*, 8 (1954): 115-118.

Lazell, J.D., Jr. "Biogeography of the Herptofauna of the British Virgin Islands, with Description of a New Anole (Sauria: Iguanidae)." Pp. 99-117 in *Advances in Herpetology and Evolutionary Biology*, edited by A.G.J. Rhodin and K. Miyata. Cambridge, Mass.: Museum of Comparative Zoology, 1983.

Le, Minh Duc. "Distribution Patterns of Reptiles and Amphibians in Puerto Rico." Master's Thesis, Oregon State University, Corvallis, Ore., 1999.

Ledesma, Hector R. "Analysis of Selected Soil Properties and Environmental Relationships in Las Cueva Watershed, Dominican Republic." Master's Thesis, University of Florida, Gainesville, Fla., 1983.

Ledesma, Hector R. "Effects of Reservoir Sedimentation and the Economics of Watershed Management: Case of Agucate Dam, Dominican Republic." Ph.D. Dissertation, University of Florida, Gainesville, Fla., 1996.

Lehmann, H. "Der tropischen Kegelkarst auf der Grossen Antillen." *Erdkunde,* 8, No. 2 (1954): 130-138.

Lehmann, H., K. Krommelbein, and W. Lotschert. "Karstmorphologischen, geologische im botanischen Studien in der Sierra de los Organos auf Cuba." *Erdkunde*, 10, No. 3 (1956): 185-203.

Leone, Frederic. "Une approche quantitative de la cartographie des risques naturels. Application experimentale au patrimoine bati de la Martinique (Antilles francaises)." *Geomorphologie*, 2 (2004): 117-126.

Leone, Frederic, and Thierry Lesales. "A GIS-Based Integrated Approach for the Assessment of Volcanic Risk: Application to Mount Pelee (Martinique, F.W.I.)." *Philippine Geographical Journal*, 47, No. 1-4 (2003): 135-150.

Lesales, Thierry. "L'ile de la Dominique. Un cas extreme d'exposition aux menaces volcaniques. Une approche-geographique de la vulnerabilite en contexte insulaire." *Bulletin de l'Association de Geographes Francais*, 81, No. 1 (2004): 92-102.

Lewis, Lawrence A. "The Relations of Hydrology and Geomorphology in a Humid Tropical Stream Basin: The Rio Grande de Monati, Puerto Rico." Ph.D. Dissertation, Northwestern University, Evanston, Ill., 1965.

Lewis, Lawrence A. "Some Fluvial Geomorphic Characteristics of the Monati Basin, Puerto Rico." *Annals of the Association of American Geographers*, 59, No. 2 (1969): 280-293.

Li, Chum. "Foraminifera: Their Distribution and Utility in the Interpretation of Carbonate Sedimentology Processes Around Grand Cayman, British West Indies." Ph.D. Dissertation, University of Alberta, Edmonton, Alberta, 1997.

Lind, Aulis O. "Coastal Landforms of Cat Island, Bahamas." Ph.D. Dissertation, University of Wisconsin, Madison, Wis., 1968.

Lind, Aulis O. *Coastal Landforms of Cat Island, Bahamas: A Study of Holocene Accretionary Topography and Sea-Level Change*. Chicago, Ill.: University of Chicago, Department of Geography, Research Paper, No. 122, 1969.

Lindskog, D.P. "From Saint Dominique to Haiti: Some Consequences of European Colonization on the Physical Environment of Hispanola." *Caribbean Geography*, 9, No. 2 (2000): 71-86.

Linhart, Y.B. "Local Biogeography of Plants on a Caribbean Atoll." *Journal of Biogeography*, 7, No. 2 (1980): 159-171.

Lips, R. "Speleogenesis on Cayman Brac, Cayman Island, British West Indies." Master's Thesis, McMaster University, Hamilton, Ontario, 1993.

Livingston, R. John. "Edge Forest Dynamics Along Roads in Virgin Islands National Park, St. John, U.S. Virgin Islands." Master's Thesis, University of Kansas, Lawrence, Kans., 1995.

Lobeck, Armin K. "The Physiography of Puerto Rico." Ph.D. Dissertation, Columbia University, New York, N.Y., 1917.

Lone, Chad S. "Late Holocene Environmental History in the Northeastern Caribbean: Multi-Proxy Evidence from Two Small Lakes on the Southern Slope of the Cordillera Central, Dominican Republic." Ph.D. Dissertation, University of Tennessee, Knoxville, Tenn., 2007.

Lopez-Marrero, Tania. "The Study of Land Cover Change in a Caribbean Landscape: What Has Happened in Puerto Rico During the Last Two Decades?" *Caribbean Studies*, 31, No. 2 (2003): 5-36.

Lopez-Marrero, Tania et al. "The Effect of Land Use on Soil Erosion in the Guadiana Watershed in Puerto Rico." *Caribbean Journal of Science*, 34, No. 3-4 (1998): 298-307.

MacGillibray, C.M.I. "A Review of Quantitative Techniques for Measuring: Karst Terrain, with Examples from the Caribbean." *Caribbean Geography*, 8, No. 2 (1997): 81-95.

Maharaj, Russell J. "Vegetation and Slope Stability: Examples from Upper St. Andrew, Jamaica." *Caribbean Geography*, 4, No. 1 (1993): 15-27.

Manent, Manuel B. "Zonificacion ecologo-pedogeoquimica de la ciudad de Moa en el Noreste de Cuba." *Revista Geografica Venezolana*, 35 (1994): 119-132.

Martin, P. et al. "Tropical Montane Forest Ecotones. "Climatic Gradients, Natural Disturbance, and Vegetation Zonation in the Cordillera Central, Dominican Republic." *Journal of Biogeography*, 34, No. 10 (2007): 1792-1806.

Mathys, Blake, and Julie Lockwood. "Rapid Evolution of Great Kiskadus on Bermuda: An Assessment of the Ability of the Island Rule to Predict the Direction of Contemporary Evolution in Exotic Vertebrates." *Journal of Biogeography*, 36, No. 12 (2009): 2204-2211.

May, T. "Bergwalder in der Dominikanischen Republik." *Geographische Rundschau*, 49, No. 11 (1997): 662-667.

May, T., and A. Drescher. "Der Zorn des Herrn der Winde. Der Hurrikan Georges in der Dominikanischen Republik Mittelfristige aus wirkungen in landlichen Raum." Pp. 127-144 in *Aktuelle Beitrage zur Angewandten physischen Geographie der Tropen, Subtropen, und der Regio Trirhena*, edited by R. Mackel and G. Zollinger. Freiburg, Germany: Freiburger Geographische, Hefte 60, 2000.

McClean, R. "Mechanical and Biological Erosion on Beach Rock in Barbados." Ph.D. Dissertation, McGill University, Montreal, Quebec, 1965.

McDald, M.A. "The Effects of Forest Clearance on Soil Conservation: Preliminary Findings from the Yallahs Valley, Jamaican Blue Mountains." *Caribbean Geography*, 3, No. 4 (1992): 253-259.

McFadden, B.J. "Rafting Mammals or Drifting Islands? Biogeography of the Greater Antillean Insectivores Neosphantes and Solendan." *Journal of Biogeography*, 7, No. 1 (1980): 11-22.

McFarlane, Donald A., and J. Lundberg. "A Middle Pleistocene Age and Biogeography for the Extinct Rodent *Megalomys curazensis* from Curucao, Netherlands Antilles." *Caribbean Journal of Science*, 38, No. 3-4 (2002): 278-281.

McFarlane, Donald A., Ross D.E. MacPhee, and Derek C. Ford. "Body Size Variability and a Sangamonian Extinction Model for Amblyrhiza, A West Indian Megafaunal Rodent." *Quaternary Research*, 50, No. 1 (1998): 80-89.

McGregor, Duncan. "An Investigation of Soil Status and Land Use on a Steeply Sloping Hillside, Blue Mountains, Jamaica." *Singapore Journal of Tropical Geography*, 9, No. 1 (1989): 60-71.

McGregor, Duncan. "Soil Erosion, Environmental Change, and Development in the Caribbean: A Deepening Crisis?" Pp. 189-208 in *Environment and Development in the Caribbean: Geographical Perspectives*, edited by David Barker and Duncan McGregor. Mona, Jamaica: University of West Indies Press, 1995.

McGregor, Duncan, and Robert B. Potter. "Environmental Change and Sustainability in the Caribbean: Terrestrial Perspectives." Pp. 1-15 in *Land, Sea, and Human Effort in the Caribbean*, edited by B. Ratter and W. Sahr. Hamburg, Germany: Beitrage zur Geographischen Regionalforschung in Lateinamerika, Volume 10, 1997.

McLaren, S.J. *Vadose Diagenesis in Late Quaternary Coastal Deposits from the Caribbean and Mediterranean: The Influence of Sea Spray*. Leicester, England: University of Leicester, Department of Geography, Occasional Paper No. 26, 1993.

Meghu, Joyse. "Some Aspects of the Physiography of Trinidad and Their Relations to Its Economic Geography." Master's Thesis, University of Toronto, Toronto, Ontario, 1957.

Meyers, Udell. "Geography of the Havana Strath." *Bulletin, Illinois Geographical Society*, 4, No. 1 (1962): 26-33.

Middleton, Nick. "Environment Today: Natural Hazards in Dominica." *Geography Review*, 22, No. 4 (2009): 18-21.

Miller, David J. "Invasion of the Cockpits: Patterns of Encroachment into the Wet Limestone Forest of Cockpit Country, Jamaica." Pp. 373-389 in *Resource Sustainability in Caribbean Development*, edited by Duncan McGregor et al. Mona, Jamaica: University of West Indies Press, 1998.

Molinelli, Ian. "Morphometric Changes in the Stream Drainage Network at the Rio Piedras River Basin in San Juan, Puerto Rico: The Impact of Urban Sprawl." Master's Thesis, State University of New York, Buffalo, N.Y., 2006.

Morales, Bolivar T. "La Biodiversidad y su Papel en el Desarrollo del Ecoturismo." *Revista Geografica*, 124 (1998): 21-32.

Morisset, Jean. "Microcosmes a la derive de l'histoire. Organisation et activites de quatre bourgs des Petites Antilles. Etude geographique." Master's Thesis, University of Laval, Ste. Foy, Quebec, 1968.

Morrison, Lloyd W. "A Review of Bahamian Ant (*Hymenoptera: Formicidae*) Biogeography." *Journal of Biogeography*, 25, No. 3 (1998): 561-572.

Morrison, Lloyd W. "Island Biogeography and Metapopulation Dynamics of Bahamian Ants." *Journal of Biogeography*, 29, No. 3 (2002): 387-394.

Morrison, Lloyd W. "Determinants of Plant Species Richness on Small Bahamian Islands." *Journal of Biogeography*, 29, No. 7 (2002): 931-942.

Morrison, Lloyd W. "Disequilibriel Island Turnover Dynamics: A 17-Year Record of Bahamian Ants." *Journal of Biogeography*, 37, No. 11 (2010): 2148-2157.

Morrison, Lloyd W., and David Spiller. "Land Hermit Crab (Coenobita clypeata) Densities and Patterns of Gastropod Shell Use on Small Bahamian Islands." *Journal of Biogeography*, 33, No. 2 (2006): 314-322.

Morrison, Lloyd W., and David Spiller. "Patterns and Processes in Insular Floras Affected by Hurricanes." *Journal of Biogeography*, 35, No. 9 (2008): 1701-1710.

Mujica-Ortiz, Banery, and Michael J. Day. "Karst Conservation and Protected Areas in Northern Puerto Rico." *Caribbean Geography*, 12, No. 1 (2001): 11-23.

Naprstek, V. "Morphology of the Shelf in the Area of Guanabo, Cuba." *Acta Universitatis Carolinae, Geographica*, 5, No. 2 (1970): 33-56.

Naughton, Patrick W. "Some Aspects of Rainfall for Jamaica." *Revista Geografica*, 96 (1982): 60-66.

Naughton, Patrick W. "Flood and Landslip Damage-Repair Cost Correlations for Kingston, Jamaica." *Caribbean Geography*, 1, No. 3 (1984): 198-202.

Naughton, Patrick W. "Storm Surge Risk Problems for Kingston, Jamaica." *Revista Geografica*, 99 (1984): 93-98.

Newton, Ella C. "Me Island Have Mighty Fine Jungle: Contentions Over the Cloud Forest of Montserrat." Master's Thesis, University of Tennessee, Knoxville, Tenn., 1999.

Nicholas, K.E. et al. "Mainland Colonization by Island Lizards." *Journal of Biogeography*, 32, No. 6 (2005): 929-938.

Niekisch, M., and A. Wezel. "Schutzgebiete in Kuba. Entwicklung und Problem." *Natur und Landschaft*, 78, No. 8 (2003): 360-366.

Nkemdirim, Lawrence C. "Spatial and Seasonal Distribution of Rainfall and Runoff in Jamaica." *Geographical Review*, 69, No. 3 (1979): 288-301.

Nkemdirim, Lawrence C. "Response Options to Sea Level Change in the Caribbean." Pp. 60-69 in *Climate, Climate Change, and Impacts*. The Hague, Netherlands: International Geographical Union, Publication of the Commission on Climatology, 1996.

Nkemdirim, Lawrence C. "Climate and Societies in the Caribbean Basin." Pp. 265-386 in *Climate and Societies*, edited by M. Yoshino et al. Dordrecht, Netherlands: Kluwer, 1997.

Nunez Jimenez, A. "On the Cuban Karst Landscapes (in Czech)." *Acta Universitatis Carolinae, Geographica*, 2 (1967): 27-47.

Obid, Jamal. "Paleomagnetic Correlations of Paleosols on San Salvador Island, Bahamas." Master's Thesis, Mississippi State University, College Station, Miss., 2000.

Oduber, Marck. "El Nino and La Nina Teleconnections on the ABC (Aruba, Bonaire, Curacao) Islands." Master's Thesis, Mississippi State University, College Station, Miss., 2000.

Oguntoyinbo, J.S. "Evapotranspiration and Sugar-Cane Yields in Barbados." *Journal of Tropical Geography*, 22 (1966): 38-48.

Oneal, Elen et al. "Testing for Biogeographic Mechanisms Promoting Divergence in Caribbean Crickets (Genus Amphiacusta)." *Journal of Biogeography*, 37, No. 3 (2010): 530-540.

Ormrod, Richard K. "The Evolution of Soil Management Practices in Early Jamaican Sugar Planting." *Journal of Historical Geography*, 5, No. 2 (1979): 157-170.

Orvis, Kenneth H. et al. "Geomorphic Traces of Quaternary Climates in the Cordillera Central, Dominican Republic." *Mountain Research and Development*, 17, No. 4 (1997): 323-331.

Pagney, Francoise. "Un cyclone dans le mer Caraibe: Le cyclone Janet (Sept. 1955)." *Les Cahiers d'Outre Mer*, 10 (1957): 65-91.

Pagney, Francoise. "Genese et dynamique de l'ouragan Hugo sur la Guadeloupe." *Annales de Geographie*, 100 (1991): 152-165.

Pagney, Francoise. "Villes de piedmont a risques d'inondation en iles tropicales: exemple des Antilles Francais." *Revue de geographie alpine*, 82, No. 4 (1994): 45-67.

Pagney, P. "Mouvements marins d'origine cyclonique. Leurs manifestations dans la Mer Caraibe specialement sur les cotes de la Martinques." *Bulletin de la Association Geographes Francais*, 276-277 (1958): 61-72.

Palm, Risa, and Michael E. Hodgson. "Natural Hazards in Puerto Rico." *Geographical Review*, 83, No. 3 (1993): 280-289.

Panos, V., and O. Steiel. "Physiogeographic and Geologic Control in Development of Cuban Mogotes." *Zeitschrift fur Geomorphologie*, 12 (1968): 117-165.

Patel, Fatima. "Coastal Development and Geomorphological Process: Scotland District, Barbados." Pp. 209-232 in *Environment and Development in the Caribbean: Geographical Perspectives*, edited by David Barker and Duncan McGregor. Kingston, Jamaica: University of the West Indies Press, 1995.

Peebles, Truman F. "A Factor Analysis and Hierarchial Grouping of Eighty-Eight Soil Profiles of Puerto Rico and the Virgin Islands." Master's Thesis, Columbia University, New York, N.Y., 1970.

Peeters, J. "Erosion et glissements de terrain a la Barbade." *Revue Belge de Geographie*, 87, No. 2 (1963): 211-226.

Peros, Matthew C. "Middle to Late Holocene Environmental Change and Archaeology on the North Coast of Central Cuba." Ph.D. Dissertation, University of Toronto, Toronto, Ontario, 2005.

Poly, A. "A Chronological Table Comprising 400 Cyclonic Hurricanes Which Have Occurred in the West Indies and in the North Atlantic Within 362 Years, from 1493-1855." *Journal of the Royal Geographical Society*, 25 (1855): 291-328.

Pulina, M., and J.R. Fagundo. "Tropical Karst and Chemical Denudation of Western Cuba." *Geographia Polonica*, 60 (1992): 195-216.

Radenbaugh, Todd A., and Adrian Seaborne. "The Status of Plant Communities in the Duncans Bay Area on Jamaica's North Coast." *Caribbean Geography*, 7, No. 2 (1996): 97-112.

Radtke, Ulrich. *Marine Terrasssen und Korallenriffe: Das Problem der quarteren Meeresspiegelschwankungen erlautert an Fallstudien aus Chile, Argentinien, und Barbados*. Dusseldorf, Germany: Dusseldorfer Geographische Schriften, Volume 24, 1989.

Radtke, Ulrich, and Gerhard Schellmann. "Uplift History Along the Clermont Nose Traverse on the West Coast of Barbados During the Last 500,000 Years: Implications for Paleo-Sea Level." *Journal of Coastal Research*, 22, No. 2 (2006): 350-360.

Randall, R.E. "Vegetation Zones and Environment of the Barbados Coast." Master's Thesis, McGill University, Montreal, Quebec, 1968.

Randall, R.E. "Measurements of Aerial Salt on the Coast of Barbados." *Climatological Bulletin*, 3 (1968): 23-35.

Raphael, C.N. "Coastal Morphology: Southwest Great Abaco Island, Bahamas." *Geoforum*, 6, No. 3-4 (1975): 237-246.

Reading, Alison J., and Rory P.D. Walsh. "Tropical Cyclone Activity within the Caribbean Basin Since 1500." Pp. 124-146 in *Environment and Development in the Caribbean: Geographical Perspectives*, edited by David Barker and Duncan McGregor. Kingston, Jamaica: University of the West Indies Press, 1995.

Reed, W.W. "Climatological Data for the West Indian Islands." *Monthly Weather Review*, 24 (1926): 133-160.

Revert, E. "La Montagne Pelee et ses derniers eruptions." *Annales de Geographie*, 40 (1931): 275-291.

Richards, A., and J.B. Bird. *Beach Studies on the West Coast of Barbados, 1967-1969: A Report*. Montreal, Quebec: McGill University, Department of Geography, 1970. 83 pp.

Richardson, H.C. "Hurricane-Borne African Locusts (*Schistocerca gregoria*) on the Windward Islands." *GeoJournal*, 23, No. 4 (1991): 349-360.

Richardson, J. "Some Implications of Tropical Forest Replacement in Jamaica." *Zeitschrift fur Geomorphologie*, Supplement, 44 (1982): 107-118.

Robertson, Richard E., Lloyd Lynch, and Joan Catchman. "Volcano Surveillance and Hazard Mitigation in the Eastern Caribbean." *Caribbean Geography*, 8, No. 1 (1997): 1-17.

Rodgers, John C. "The Distribution of Casuarians on San Salvador Island, the Bahamas." *Southeastern Geographer*, 45, No. 2 (2005): 222-238.

Rodriguez, Mario G.C. "Estudio geomorfologico del carso en la region Gibara-Banes de Holguin, Cuba." *Revista Geografica*, 134 (2003): 5-22.

Rodriguez, Mario G.C., and Diego C. Barba. "The Hispanola Fluvial System and Its Morphostructural Context." *Physical Geography*, 30, No. 5 (2009): 453-478.

Rodriguez, Mario G.C. et al. "La red fluvial de Cuba y su interpretacion morfoestructural." *Revista Geografica*, 134 (2003): 23-50.

Rodrguez-Paneque, Ridel, and Elier A. Cordova Garcia. "La erosion en las playas de la region nororiental de Cuba." *Revista Geografica*, 139 (2006): 9-26.

Rosen, D.E. "A Vicariance Model of Caribbean Biogeography." *Systematic Zoology*, 24 (1975): 431-468.

Roth, Linda. "Traditional Knowledge and Management of Subtropical Dry Forest: A Study of Resource Use, Landscape Change, and Biodiversity in a Rural Area of the Dominican Republic." Master's Thesis, Clark University, Worcester, Mass., 1997.

Roth, Linda. "Anthropogenic Change in Subtropical Dry Forest During a Century of Settlement in Jaiqui Picado, Santiago Province, Dominican Republic." *Journal of Biogeography*, 26, No. 4 (1999): 739-760.

Roth, Monica. "Inventory and Geometric Analysis of Plank Margin Caves of the Bahamas." Master's Thesis, Mississippi State University, State College, Miss., 2004.

Rouse, Wayne R. *The Moisture Balance of Barbados and Its Influence on Sugar Cane Yield*. Montreal, Quebec: McGill University, Department of Geography, Climate Research Series No. 1, 1966.

Rouse, Wayne R., and David J. Watts. *Two Studies in Barbadian Climatology*. Montreal, Quebec: McGill University, Climatological Research Series, No. 1, 1966.

Rudd, Thomas K., Marla Perez-Hugo, and Heather Zichal. "When Fields Revert to Forest: Development and Spontaneous Reforestation in Post-War Puerto Rico." *Professional Geographer*, 52, No. 3 (2000): 386-396.

Salomon, J. "Relation entre karst, aquifers et niveaux de la mer a Cuba." *Hommes et Terres du Nord*, 1 (1995): 82-96.

Sauer, Jonathan D. *Cayman Island Seashore Vegetation: A Study in Comparative Biogeography*. Berkeley, Calif.: University of California Press, 1982.

Scatena, F.N., and Ariel E. Lugo. "Geomorphology, Disturbance, and the Soil and Vegetation of Two Subtropical Wet Steepland Watersheds of Puerto Rico." *Geomorphology*, 13, No. 1-4 (1995): 199-213.

Scheffers, A. *Coastal Response to Extreme Wave Event: Hurricanes and Tsunami on Bonaire*. Essen, Germany: Essener Geographische Arbeiten, Volume 37, 2005.

Scheffers, A., and S. Scheffers. "Documentation of the Impact of Hurricane Ivon on the Coastline of Bonaire (Netherlands Antilles)." *Journal of Coastal Research*, 22, No. 6 (2006): 1437-1450.

Scheffers, A. et al. "Paleo-Tsunami Relics on the Southern and Central Antillean Island Arc." *Journal of Coastal Research*, 21, No. 2 (2005): 263-273.

Schellmann, Gerhard, and Ulrich Radtke. *The Marine Quaternary of Barbados*. Koln, Germany: Kolner Geographische Arbeiten, Band 81, 2004.

Schellmann, Gerhard et al. "Erythcocyte Sedimentation Rate (ESR) Dating of Coral Reef Terraces on Curacao (Netherlands Antilles) with Estimates of Younger Pleistocene Sea Level Elevations." *Journal of Coastal Research*, 20, No. 4 (2004): 947-957.

Schiettecatte, W. et al. "Influence of Land Use on Soil Erosion Risk in the Cuyaguateje Watershed (Cuba)." *Catena*, 74, No. 1 (2008): 1-12.

Schubert, Carlos. "Investigaciones sobre el Cuaternario de la Republica Dominicana." *Revista Geografica*, 99 (1984): 69-92.

Scruggs, Paul Delmas. "Initial Results of In Situ Infiltration Rates in the Luquillo Experimental Forest, Puerto Rico, and Recommendations for Future Research." Master's Thesis, University of Tennessee, Knoxville, Tenn., 2003.

Seal, J.N. et al. "Phylogeography of the Parthenogenic Ant *Platythyrea punctata*: Highly Successful Colonization of the West Indies by a Poor Disperser." *Journal of Biogeography*, 38, No. 5 (2011): 868-882.

Sealey, Neil E. "Bahamian Blue Holes: Windows on the Past, Resources for the Future." Pp. 35-45 in *Environment and Development in the Caribbean: Geographical Perspectives*, edited by David Barker and Duncan McGregor. Kingston, Jamaica: University of the West Indies Press, 1995.

Sebastian, Glenn R. "Cayman Brac: A Micro Study in Geographic Plant Distribution." *Mississippi Geographer*, 3, No. 1 (1975): 67-74.

Seguinot, Jose B. "Coastal Modification and Land Transformation in the San Juan Bay Area, Puerto Rico." Ph.D. Dissertation, Louisiana State University, Baton Rouge, La., 1983.

Shepard, J.B. "Earthquake and Volcanic Hazard Assessment and Monitoring in the Commonwealth Caribbean: Current States and Needs for the Future." Pp. 50-60 in *Proceedings of a Meeting of Experts on Hazard Mapping in the Caribbean*, edited by David Barker. Mona,

Jamaica: University of the West Indies, Department of Geography, 1989.

Shreve, F. *A Montane Rain-Forest: A Contribution to the Physiological Plant Geography of Jamaica.* Washington, D.C.: Carnegie Institute, Publication No. 199, 1914.

Simon, A., Matthew C. Larsen, and C.R. Hupp. "The Role of Soil Processes in Determining Mechanisms of Slope Failure and Hillslope Development in a Humid-Tropical Forest, Eastern Puerto Rico." *Geomorphology*, 3, No. 3-4 (1990): 263-286.

Simonti, Alicia L. "The 2005 Caribbean Mass Coral Bleaching Event: A Sea Surface Temperature Empirical Orthoganal Teleconnection Analysis." Master's Thesis, Clark University, Worcester, Mass., 2008.

Simpson, George G. *Zoogeography of West Indian Land Mammals.* New York, N.Y.: American Museum Novitates, No. 1759, 1956.

Singh, B. "Climate-Induced Global Changes in the Southern Caribbean: Trinidad and Tobago." *Global Planetary Changes*, 15 (1997): 93-111.

Singh, B. "Climatic Changes in the Greater and Southern Caribbean." *International Journal of Climatology*, 17 (1997): 1093-1114.

Singh, B. "Greenhouse Gas Emissions and Constraints in the Small Island States of the Caribbean." Pp. 313-326 in *Air Pollution*, edited by C.A. Brebbin and J.F. Martin-Duque. Southampton, England: WIT Press, 2002.

Smart, C.C. "The Hydrology of the Inland Blue Holes, Andros Island." *Cave Science*, 11, No. 1 (1984): 23-29.

Smith, D.I., D.P. Drew, and T.C. Atkinson. "Hypothesis of Karst Landform Development in Jamaica." *Transactions, Cave Research Group*, 14 (1972): 159-173.

Smith, Jason K. "An Assessment of Blue Mahoe (*Hibiscus elatus*) Growth Using Topographic, Soil, and Litter Properties in St. Vincent, West Indies." Master's Thesis, University of Calgary, Calgary, Alberta, 2001.

Smith, Samuel I. "An Analysis of Rainfall and Sugarcane Yields in Barbados, West Indies, 1937-1958." *Bulletin, Illinois Geographical Society*, 17, No. 1 (1975): 14-31.

Soto, Sandra A. "Solving a Locational Distribution Problem of Non-Toxic Solid Waste on the Island of Puerto Rico." Master's Thesis, Louisiana State University, Baton Rouge, La., 2004.

Soto, Sandra A. and Josep Pinto. "Delineation of Natural Landscape Units for Puerto Rico." *Applied Geography*, 30, No. 4 (2010): 720-730.

Speer, James et al. "Assessing the Dendrochronological Potential of *Pinus occidentalis Swartz* in the Cordillera Central of the Dominican Republic." *The Holocene*, 14, No. 4 (2004): 563-569.

Steers, J.A. "The Cays and the Palisades, Port Royal, Jamaica." *Geographical Review*, 30 (1940): 279-296.

Steers, J.A. "The Coral Cays of Jamaica." *Geographical Journal*, 95 (1940): 30-42.

Steers, J.A. et al. "Sand Cays and Mangroves in Jamaica." *Geographical Journal*, 96 (1940): 305-328.

Stehle, H. *Flore de la Guadeloupe et Depenances. Essai d'Ecologie et de Geographie botanique*. Basse-Terre, Guadeloupe: Imprimerie Officielle de la Martinique, 1939.

Stork, Allison J. "A Paleoecological History of West Pond on Great Abaco Island, the Bahamas, Based on Pollen and Charcoal Analyses of Lake Sediments." Master's Thesis, University of Tennessee, Knoxville, Tenn., 2006.

Strand, James T. "Soil Erosion in a Blue Mahoe (*Hibiscus natus*) Forest Plantation and a Secondary Rainforest of the Colonarie River Watershed, St. Vincent, West Indies." Master's Thesis, University of California, Los Angeles, Calif., 1996.

Stuart, Kimberly. "The Potential of Wind Energy for the Operation of Reverse Osmosis Water Plants in the Bahamas." Master's Thesis, University of Nebraska, Omaha, Neb., 2009.

Sweeting, Marjorie M. "Notes on the Caves of Jamaica." *Transactions of the Caves Research Group of Great Britain*, 5 (1957): 1-11.

Sweeting, Marjorie M. "The Karstlands of Jamaica." *Geographical Journal*, 124 (1958): 184-199.

Tam, Sai Wing. "Mechanisms and Spatial Patterns of Erosion and Instability in the Joe's River Basin, Barbados." Ph.D. Dissertation, McGill University, Montreal, Quebec, 1976.

Terborh, J. “Chance, Habitat, and Dispersal in the Distribution of Birds in the West Indies.” *Evolution*, 27 (1973): 338-349.

Torres, Rebecca M. “Introduction: Globalization, Development, and Environmental Change in the Caribbean.” *Southeastern Geographer*, 45, No. 2 (2005): 165-173.

Tout, D.G. “The Climate of Barbados.” *Climatological Bulletin*, 3 (1968): 1-17.

Trejo-Torres, J.C., and J.D. Ackerman. “Biogeography of the Antilles Based on a Parsimony Analysis of Orchid Distribution.” *Journal of Biogeography*, 28, No. 6 (2001): 775-794.

Urquhart, A W. *The Landforms of the Cockpit Country and Its Borderlands*. Berkeley, Calif.: University of California, Department of Geography, 1958.

Valdes, Angel, and Pedro Rodriguez. “El componente biotico y su papel como indicadar de las paisajes en el ejenplo del curso inferior del rio Jaguari, Cuba.” *Revista Geografica Venezolana*, 35 (1994): 109-11.

Van Bloem, Skip et al. “Structural Response of Caribbean Dry Forests to Hurricane Winds: A Case Study from Guavica Forest, Puerto Rico.” *Journal of Biogeography*, 33, No. 3 (2006): 517-523.

Vermette, Stephen, and Rebecca Hudson. “Hydrology of Walting’s Blue Hole: San Salvador, Bahamas.” *Middle States Geographer*, 34 (2001): 55-62.

Versey, H. “Karst of Jamaica.” Pp. 445-466 in *Karst: Important Karst Regions of the Northern Hemisphere*, edited by Milan Herak. New York, N.Y.: Elsevier, 1972.

Visser, Lloyd L. “Geographic Aspects of P-Fixation on St. Vincent, West Indies.” Master’s Thesis, University of Calgary, Calgary, Alberta, 1989.

Vogt, J. “Revision de deux seimes des Petites Antilles: 1839 et 1851.” *Revue de Geomorphologie Dynamique*, 43, No. 4 (1994): 135-143.

Waibel, Leo. “Place Names as an Aid in the Reconstruction of the Original Vegetation of Cuba.” *Geographical Review*, 33 (1943): 376-396.

Walsh, Rory P.D. “Changes in the Tracks and Frequencies of Tropical Cyclones in the Lesser Antilles from 1560 to 1975 and Some Geomorphological and Ecological Implications.” *Swansea Geographer*, 15 (1977): 4-11.

Walsh, Rory P.D. "Drainage Density and Hydrological Process in a Humid Tropical Environment: The Windward Islands." Ph.D. Dissertation, University of Cambridge, Cambridge, England, 1980.

Walsh, Rory P.D. "Climatic Changes in the Eastern Caribbean Over the Past 150 Years and Some Implications in Planning Sustainable Development." Pp. 26-50 in *Resource Sustainability and Caribbean Development*, edited by Duncan McGregor et al. Mona, Jamaica: University of West Indies Press, 1998.

Walsh, Rory P.D., and K. Howells. "Soil Pipes and Their Role in Runoff Generation and Chemical Denudation in a Humid Tropical Catchment in Dominica." *Earth Surface Processes and Landforms*, 13, No. 1 (1988): 9-17.

Walsh, Rory P.D., and Alison J. Reading. "Historical Changes in Tropical Cyclone Frequency within the Caribbean Since 1500." *Wurzberger Geographische Arbeiten*, 80 (1991): 199-240.

Watts, David J. *Evapotranspiration and Energy Relationships at Waterford, Barbados*. Montreal, Quebec: McGill University, Department of Geography, Climate Research Series, No. 1, 1966.

Watts, David J. *Man's Influence on the Vegetation of Barbados, 1627 to 1800*. Hull, England: University of Hull, Occasional Paper in Geography, No. 4, 1966.

Watts, David J. "Long-Term Environmental Influences on Development in Islands of the Lesser Antilles." *Scottish Geographical Magazine*, 109, No. 3 (1993): 133-141.

Webster, Edgar H. "Tides of the Gulf of Mexico and the Caribbean Sea." *Bulletin of the Geographical Society of Philadelphia*, 30 (1932): 214-224.

Wesma, Hector P. "Effects of Reservoir Sedimentation and the Economics of Watershed Management: Case of Agucate Dam, Dominican Republic." Ph.D. Dissertation, University of Florida, Gainesville, Fla., 1996.

Whitaker, Fiona F., and Peter L. Smart. "Groundwater Circulation and Geochemistry of a Karstfield Bank: Marginal Fracture System, South Andros Island, Bahamas." *Journal of Hydrology*, 197 (1997): 293-315.

Wills, Bernt L. "The Limestone Belt of the North Central Area of Puerto Rico." Pp. 71-96 in *Symposium on the Geography of Puerto Rico*, edited by Clarence Jones and Rafael Pico. Rio Piedres, Puerto Rico: University of Puerto Rico Press, 1955.

Wilson, Herbert M. "Porto Rico: Its Topography and Aspects." *Journal of the American Geographical Society of New York*, 32, No. 3 (1900): 220-238.

Wilson, Jeffrey S., Timothy S. Brothers, and Eugenio Marcano. "Land Cover Contrasts Along the Haitian-Dominican Republic Border." *Caribbean Geography*, 11, No. 1 (2000): 7-19.

Wilson, Mark. *The Caribbean Environment*, Second Edition. London, England: Oxford University Press, 1997.

Wilson, Peter. "Comments on a Possible Relationship Between Differential Radiation Balances and Rainfall Patterns in Barbados." *Climatological Bulletin*, 27 (1980): 19-30.

Wong, Poh Poh. "Beach Changes and Sand Movement in Low Energy Environments, West Coast, Barbados." Ph.D. Dissertation, McGill University, Montreal, Quebec, 1971.

Wood, Harold A. "Stream Piracy in the Central Plateau of Hispanola." *Canadian Geographer*, 8 (1956): 46-54.

Wood, P.A. "Suspended Sediment in a Tropical Environment of Seasonal Flow and Large Floods: Hope River, Jamaica." *Journal of Tropical Geography*, 45 (1977): 65-69.

Woodroffe, C.D., D.R. Stoddart, and M.E.C. Giglioli. "Pleistocene Patch Reefs and Holocene Swamp Morphology, Grand Cayman Island, West Indies." *Journal of Biogeography*, 7, No. 2 (1980): 103-113.

Woodroffe, C.D. et al. "Coastal Morphology and Late Quaternary History, Cayman Islands, West Indies." *Quaternary Research*, 19, No. 1 (1983): 64-84.

Woods, C.A., ed. *Biogeography of the West Indies, Past, Present and Future*. New York, N.Y.: Academic Press, 1978.

Young, Robert N. "A Geographic Classification of Landform Types of Puerto Rico." Pp. 27-46 in *Symposium on the Geography of Puerto Rico*, edited by Clarence Jones and Rafael Pico. Rio Piedras, Puerto Rico: University of Puerto Rico Press, 1955.

PART 5

Historical Geography

Adams, John E. "Historical Geography of Whaling in Bequia Island, West Indies." *Caribbean Studies*, 11 (1971): 55-74.

Adams, John E. "Union Island, West Indies: An Historical and Geographical Sketch." *Caribbean Studies*, 18, No. 2 (1978): 5-45.

Adams, John E. "Last of the Caribbean Whalemen." *Natural History*, 103, No. 11 (1994): 64-72.

Ashdown, P. *Caribbean History in Maps*. London, England: Longman, 1979.

Berry, M. Victoria. "Geographic Origins of Medicinal Plants in Montserrat, West Indies, Pre-1632 to Present." *Southeastern Geographer*, 43, No. 1 (2003): 75-88.

Bowie, Jean M. "An Historical Geography of the Island of Trinidad." Master's Thesis, University of California, Berkeley, Calif., 1951.

Clarke, Colin G. "A Caribbean Creole Capital: Kingston, Jamaica 1692-1938." Pp. 153-170 in *Colonial Cities*, edited by Robert J. Ross and Gerard J. Telkamp. Dordrecht, Netherlands: Maritimes Nijhof, 1985.

Clarke, Colin G., and David Howard. "Color, Race, and Space: Residential Segregation in Kingston, Jamaica, in the Late Colonial Period." *Caribbean Geography*, 10, No. 1 (1999): 4-18.

Cline, H.F. "The Relationes Geograficos of the Spanish Indies, 1577-1586." *Hispanic American Historical Review*, 44 (1964): 341-374.

Crist, Raymond E. "History of the Coffee Industry in Cuba." *Geographical Review*, 36 (1946): 493-495.

Day, Michael J. "Military Campaigns in Tropical Karst: The Maroon Wars of Jamaica." Pp. 79-88 in *Studies in Military Geography and Geology*, edited by Douglas R. Caldwell, Judy Ehlen, and Russell S. Harmon. Dordrecht, Netherlands: Kluwer Academic, 2004.

Demeritt, David. "Board, Barrels, and Boxshooks: The Economies of Downeast Lumber in Nineteenth-Century Cuba." *Forest and Conservation History*, 35, No. 3 (1991): 108-120.

Dierickx, Charles W. "An Historical Geography of the Bahamas Islands." Master's Thesis, University of Florida, Gainesville, Fla., 1952.

DiRazza, N. "Les nouveaux defies du processus Onusien de consolidation de la paix en Haiti." *Bulletin de l'Association de Geographes Francais*, 87, No. 3 (2010): 293-307.

Doran, Edwin, Jr. "This Columbus-Caicos Confusion." *Professional Geographer*, 13 (1961): 32-34.

Elbow, Gary S. "Migration or Interaction: Reinterpreting Pre-Columbian West Indian Culture Origins." *Journal of Geography*, 91, No. 5 (1992): 200-204.

Euwena, Jeffrey A. "The Decline of Sugar Production and the Rise of Cocoa Production in Grenada, 1870-1917: The Changing Fortunes of a Cocoa Peasantry." Master's Thesis, Virginia Polytechnic Institute and State University, Blacksburg, Va., 1993.

Everad, Kyle T. "Fostering Puerto Rico: Representation of Empire and Orphaned Territories During the Spanish-American War." *Historical Geography*, 34 (2006): 109-136.

Eyre, L. Alan. "The Maroon Wars in Jamaica: A Geographical Appraisal." *Jamaica Historical Review*, 12 (1980): 80-102.

Fentem, Arlin. "The Historical Geography of a Tropical Economy-Dominica, T.W.I." *Bulletin, Illinois Geographical Society*, 2, No. 1 (1960): 4-15.

Fuson, Robert. "Caicos: Site of Columbus' Landfall." *Professional Geographe,* 13 (1961): 6-9.

Fuson, Robert. "Caicos, Confusion, Conclusion." *Professional Geographer*, 13 (1961): 35-37.

Galloway, J.H. "The Settlement of Barbados and the Establishment of the Plantation Economy, 1627-1680." Master's Thesis, University of California, Berkeley, Calif., 1961.

Galloway, J.H. "The Sugar Industry in Barbados During the Seventeenth Century." *Journal of Tropical Geography*, 19 (1964): 35-41.

Galloway, J.H. "Botany in the Service of Empire: The Barbados Cane-Breeding Program and the Revival of the Caribbean Sugar Industry, 1880s-1930s." *Annals of the Association of American Geographers*, 86, No. 4 (1996): 682-706.

Galloway, J.H. "Abolition and After: The Nineteenth Century Crisis in the Sugarcane Industry." Pp. 27-38 in *Slaves With or Without Sugar*, edited by A. Vieira. Madeira, Portugal: Center for the Study of Atlantic History, 1996.

Galvin, Peter R. *Patterns of Pillage: A Geography of Caribbean-Based Piracy in Spanish America, 1536-1718*. New York, N.Y.: Peter Lang, 1999.

Gibbes, Cerian, Jane Southworth, and Eric Keys. "Wetland Conservation: Change and Fragmentation in Trinidad's Protected Areas." *Geoforum*, 40, No. 1 (2008): 91-104.

Gordon, Sadie C.B. "Geographic Influences in the History of Haiti." Master's Thesis, Columbia University, New York, N.Y., 1937.

Grossman, Lawrence S. "The Colonial Office and Soil Conservation in the British Caribbean, 1938-1950." Pp. 29-40 in *Caribbean Land and Development Revisited*, edited by J. Besson and Janet H. Momsen. New York, N.Y.: Palgrave Macmillan, 2007.

Higman, Barry W. "The Spatial Economy of Jamaican Sugar Plantations: Cartographic Evidence from the Eighteenth and Nineteenth Centuries." *Journal of Historical Geography*, 13, No. 1 (1987): 17-39.

Higman, Barry W. *Jamaica Surveyed: Plantation Maps and Plans of the Eighteenth and Nineteenth Centuries*. Jamaica: Institute of Jamaica Publications, 1988.

Higman, Barry W. *Montpelier, Jamaica: A Plantation Community in Slavery and Freedom, 1739-1912*. Mona, Jamaica: University Press of the West Indies, 1998.

Hopkins, Daniel P. "The Danish Cadastral Survey of St. Croix, 1733-1754." Ph.D. Dissertation, Louisiana State University, Baton Rouge, La., 1988.

Hopkins, Daniel P. "An Early Map and the Cadastral Survey of St. Croix, Danish West Indies, 1734-41: A Cartographic Cul-de-Sac." *Cartographica*, 29, No. 3 (1992): 1-9.

Hopkins, Daniel P. "The Eighteenth-Century Invention of a Measure in the Caribbean: The Danish Acre of St. Croix." *Journal of Historical Geography*, 18, No. 2 (1992): 158-173.

Howard, Richard A. "A History of the Botanic Garden of St. Vincent, British West Indies." *Geographical Review*, 44, No. 3 (1954): 381-393.

Hubber, L.L. "Did Columbus Discover the Islands of Antigua and St. Martin?" *Geographical Review*, 21, No. 4 (1931): 584-597.

Huetz de Lemps, C. "Deux nouveaux ouvrages sur l'histoire antillaise." *Les Cahiers d'Outre Mer*, 36 (1983): 191-195.

Huish, Robert L. "Made of Paper and Stone: The Place of Jose Marti in Cuban National Identity." Master's Thesis, Queen's University, Kingston, Ontario, 2003.

Innes, Frank C. "Plantation and Peasant Farm: A Vertical Theme in the Historical Geography of Barbados: 1627-1960." Ph.D. Dissertation, McGill University, Montreal, Quebec, 1967.

Innes, Frank C. "The Pre-Sugar Era of European Settlement in Barbados." *Journal of Caribbean History*, 1 (1970): 1-22.

Innes, Frank C. "Approaches to a Study of Small-Scale Agriculture in the Commonwealth Caribbean." Pp. 181-186 in *Proceedings, Canadian Association of Geographers*. Winnipeg, Manitoba: University of Manitoba, Department of Geography, 1970.

Innes, Frank C. "News from the Caribbean Islands: A Review Article." *Historical Geography Newsletter*, 3, No. 3 (1973): 6-10.

Innes, Frank C. "Economic History and Historical Geography of the Caribbean: The Last Ten Years Reviewed." *Proceedings, Conference of Latin Americanist Geographers* (1981): 392-396.

Jaquay, Barbara. "The Caribbean Cotton Production: An Historical Geography of the Region's Mystery Crop." Ph.D. Dissertation, Texas A&M University, College Station, Tex., 1997.

Johnson, H. "Bahamian Labor Migration to Florida in the Late Nineteenth and Early Twentieth Centuries." *International Migration Review*, 22, No. 1 (1988): 84-103.

Joubert, E. "Consequences geographique de l'emancipation des nours aux Antilles." *Les Cahiers d'Outre Mer*, 1 (1948): 103-118.

Kimber, Clarissa. "Recent Historical Plant Geography of Martinique." Ph.D. Dissertation, University of Wisconsin, Madison, Wis., 1968.

Lambert, David. "The Master Subject: White Identities and the Slavery Controversy in Barbados, 1780-1834." Ph.D. Dissertation, University of Cambridge, Cambridge, England, 2001.

Lambert, David. "Liminal Figures: Poor Whites, Foundmen, and Racial Reinscription in Colonial Barbados." *Environment and Planning D*, 19, No. 3 (2001): 334-350.

Lambert, David. "True Lovers of Religion: Methodist Persecution and White Resistance to Anti-Slavery in Barbados, 1823-1825." *Journal of Historical Geography*, 28, No. 2 (2002): 216-236.

Lambert, David. "Producing/Contesting Whiteness: Rebellion, Anti-Slavery, and Enslavement in Barbados, 1816." *Geoforum*, 36, No. 1 (2005): 29-43.

Lambert, David. "The Counter-Revolutionary Atlantic: White West Indian Petitions and Proslavery Networks." *Social and Cultural Geography*, 6, No. 3 (2005): 405-420.

Lucas, Charles P. *The West Indies: Volume II of a Historical Geography of the British Colonies*. Oxford, England: Clarendon Press, 1905.

Meggers, Betty J. "Possible Impact of the Mega-Nino Events in Precolumbian Populations in the Caribbean." Pp. 156-176 in *Ponencias del Primer Seminario de Arqueologial de Caribe*, edited by Marcio Velaz Maggiolo and Angel Caba Fuentes. Altos de Clinon: Museo Arqueologico Regional y Organizacion de Estados Americanos, 1996.

Mena, Luz-Maria. "Hybrid Modernity: Havana in the 1830s and '40s." Master's Thesis, University of California, Berkeley, Calif., 1995.

Mena, Luz-Maria. "No Common Folk: Free Blacks and Race Relations in the Early Modernization of Havana, 1830s-1840s." Ph.D. Dissertation, University of California, Berkeley, Calif., 2001.

Merrill, Gordon C. "The Historical Geography of St. Kitts and Nevis, British West Indies." Ph.D. Dissertation, University of California, Berkeley, Calif., 1957.

Merrill, Gordon C. *The Historical Geography of St. Kitts and Nevis, the West Indies*. Mexico City, Mexico: Instituto Pan Americano de Geografica e Historica, 1958.

Merrill, Gordon C. "The Historical Record of Man as an Ecological Dominant in the Lesser Antilles." *Canadian Geographer*, 3, No. 11 (1958): 17-22.

Merrill, Gordon C. *The Historical Geography of St. Kitts and Nevis, the West Indies*. Mexico City, Mexico: Pan-American Institute of Geography and History, No. 232, 1958.

Merrill, Gordon C. "The Role of the Sephardic Jews in the British Caribbean Area During the Seventeenth Century." *Caribbean Studies*, 4 (1964): 32-40.

Momsen, Janet H. "The Marginalisation of Women Farmers: Gender and Small-Scale Agriculture in Barbados, 1838-1935." *Caribbean Geography*, 6, No. 1 (1995): 52-60.

Mousnier, Mireille. *Atlas historique du patrimoine sucrier de la Martinque*. Paris, France: L'Harmattan, 1990.

Nelson, Velvet. "Landscape and Postcolonialism in British West Indies Travel Narratives, 1815-1914." Ph.D. Dissertation, Kent State University, Kent, Ohio, 2006.

Nelson, Velvet. "The Sensibility of Aesthetic Landscape Concepts in the Case of British West Indies Travel Narratives, 1815-1914." *Journal of Cultural Geography*, 26, No. 2 (2009): 149-173.

Newson, Linda A. *Aboriginal and Spanish Colonial Trinidad: A Study in Culture Contact*. London, England: Academic Press, 1976.

Niddrie, David L. "An Attempt at Planned Settlement in St. Kitts in the Early Eighteenth Century." *Caribbean Studies*, 5, No. 4 (1966).

Niddrie, David L. "Eighteenth-Century Settlement in the British Caribbean." *Transactions, Institute of British Geographers*, 40 (1966): 67-80.

Ormrod, Richard K. "Adaptation in Cultural Ecosystems: Early Nineteenth Century Jamaica." Ph.D. Dissertation, Pennsylvania State University, University Park, Pa., 1974.

Ormrod, Richard K. "The Evolution of Soil Management Practices in Early Jamaican Sugar Planting." *Journal of Historical Geography*, 5, No. 2 (1979): 157-170.

Platt, Raye R. et al. *The European Possession in the Caribbean Area.* New York, N.Y.: American Geographical Society, 1941.

Price-Mars, D.J. "Aspectos generales del sistema de explotacion rural en el Nuevo Mundo: La colonizacion de Saint-Domingue y el Dessarollo de las plantaciones en la parte occidental de la isla." *Revista Geografica*, 54 (1961): 101-120.

Pulsipher, Lydia M. "The Cultural Landscape of Montserrat, West Indies, in the Seventeenth Century: Early Environmental Consequences of British Colonial Development." Ph.D. Dissertation, Southern Illinois University, Carbondale, Ill., 1977.

Pulsipher, Lydia M. "English, Irish, and African Influences on the Landscape of Seventeenth Century Montserrat in the West Indies." *Proceedings, Middle States Division, Association of American Geographers*, 12 (1979): 5-7.

Pulsipher, Lydia M. *Seventeenth-Century Montserrat: An Environmental Impact Statement.* Norwich, England: Geo Books, 1986.

Redon, M. "Frontiere poreuse, Etat faibel: Les relations Haiti/Republique dominicaine a l'aune de la frontiere." *Bulletin de l'Association de Geographes Francais*, 87, No. 3 (2010): 308-323.

Reece, Andrew G. "Colonial Dualism in the Commonwealth Caribbean: A Study of Kingston, Jamaica." Master's Thesis, University of North Carolina, Chapel Hill, N.C., 1994.

Richardson, Benjamin F. "Jamaica in Early Maps." *Journal of Geography*, 82, No. 5 (1983): 195-204.

Richardson, Bonham C. "Livelihood in Rural Trinidad in 1900." *Annals of the Association of American Geographers*, 65, No. 2 (1975): 240-251.

Richardson, Bonham C. "Freedom and Migration in the Leeward Caribbean, 1838-1848." *Journal of Historical Geography*, 6, No. 4 (1980): 391-408.

Richardson, Bonham C. "Slavery to Freedom in the British Caribbean: Ecological Considerations." *Caribbean Geography*, 1, No. 3 (1984): 164-175.

Richardson, Bonham C. *Panama Money in Barbados, 1900-1920.* Knoxville, Tenn.: University of Tennessee Press, 1985.

Richardson, Bonham C. "Caribbean Migrations, 1838-1985." Pp. 203-228 in *The Modern Caribbean*, edited by F.W. Knight and C.A. Palmer. Chapel Hill, N.C.: University of North Carolina Press, 1989.

Richardson, Bonham C. "Human Mobility in the Windward Caribbean, 1884-1902." *Plantation Society*, 2, No. 3 (1989): 301-319.

Richardson, Bonham C. *The Caribbean in the Wider World, 1492-1992*. New York, N.Y.: Cambridge University Press, 1992.

Richardson, Bonham C. "Depression Riots and the Calling of the 1897 West India Royal Commission." *Nieuwe West-Indische Gids*, 66, No. 3-4 (1992): 169-191.

Richardson, Bonham C. "A 'Respectable' Riot: Guy Fawkes Night in St. George's, Grenada, 1885." *Journal of Caribbean History*, 27, No. 1 (1993): 21-35.

Richardson, Bonham C. "Detrimental Determinants: Applied Environmentalism as Bureaucratic Self-Interest in the Fin-de-Siecle British Caribbean." *Annals of the Association of American Geographers*, 86, No. 2 (1996): 213-234.

Richardson, Bonham C. *Economy and Environment in the Caribbean: Barbados and the Windwards in the Late 1800s*. Gainesville, Fla.: University Press of Florida, 1997.

Richardson, Bonham C. *Igniting the Caribbean's Past: Fire in British West Indian History*. Chapel Hill, N.C.: University of North Carolina Press, 2004.

Richardson, Bonham C. "The Importance of the 1897 British Royal Commission." Pp. 17-28 in *Caribbean Land and Development Revisited*, edited by J. Besson and Janet H. Momsen. New York, N.Y.: Palgrave Macmillan, 2007.

Ross, W.G. "The Historical Geography of Antigua." Master's Thesis, McGill University, Montreal, Quebec, 1968.

Sauer, Carl O. *The Early Spanish Main*. Berkeley, Calif.: University of California Press, 1966.

Sealey, Neil E. "New Developments on the Columbus Landfall Issue." *Journal of the Bahamas Historical Society*, 9 (1987): 5-13.

Sealey, Neil E. "An Examination of the Geography of Three Major Contenders for Columbus' First Landfall in 1492." *Terra Incognita*, 23 (1991): 39-50.

Seymour, S., S. Daniels, and E. Watkins. "Estate and Empire: Sir George Cornewell's Management of Moccas, Herefordshire, and La Taste, Grenada, 1771-1819." *Journal of Historical Geography*, 24, No. 3 (1998): 313-350.

Singh, S.P. "Impact of East Indian Immigrants on Trinidad's Agricultural Economy: A Historical Perspective." *National Geographical Journal of India*, 32, No. 2 (1986): 137-163.

Smallman-Raynor, Matthew, and Andrew D. Cliff. "The Spatial Dynamics of Epidemic Diseases in War and Peace: Cuba and the Insurrection Against Spain, 1895-98." *Transactions, Institute of British Geographers*, 24, No. 3 (1999): 331-352.

Smith, S.D. "Volcanic Hazard in a Slave Society: The 1812 Eruption of Mount Soufriere in St. Vincent." *Journal of Historical Geography*, 37, No. 1 (2011): 55-67.

Sparling, C. "The Historical Geography of St. Vincent." Master's Thesis, Carleton University, Ottawa, Ontario, 1970.

Stewart, Glenn. "San Salvador Island to Cuba: A Cruise in the Track of Columbus." *Geographical Review*, 21 (1931): 124-130.

Street, John M. "Historical and Economic Geography of the Southwest Peninsula of Haiti." Ph.D. Dissertation, University of California, Berkeley, Calif., 1959.

Theodat, Jean-Marie. "Haiti, quisqueya: Une double insularite: Essai de geographie historique, 1804-1915." Ph.D. Dissertation, University of Lille, Lille, France, 1998.

Valdoz, Jose J. "Paleogeographic Perspectives on the First Landfall of Columbus." *Southeastern Geographer*, 34, No. 2 (1994): 73-91.

Watlington, Francisco. "Cassava and Carrying Capacity in Aboriginal Puerto Rico: Revisiting the Taino Downfall at Conquest." *Southeastern Geographer*, 49, No. 4 (2009): 394-403.

Watts, David J. "Plant Introduction and Landscape Change in Barbados, 1625-1830." Ph.D. Dissertation, McGill University, Montreal, Quebec, 1963.

Watts, David J. *Man's Influence on the Vegetation of Barbados, 1627 to 1800*. Hull, England: University of Hull, Occasional Paper in Geography No. 4, 1966.

Watts, David J. "Origins of Barbadian Cane Hole Agriculture." *Journal of the Barbados Museum and Historical Society*, 32 (1968): 143-151.

Watts, David J. "Dung-Farming: A Seventeenth-Century Experiment in Barbadian Agricultural Improvisation." *Journal of the Barbados Museum and Historical Society*, 34 (1972): 58-63.

Watts, David J. "Cycles of Famine in Islands of Plenty: The Case of the Colonial West Indies in the Pre-Emancipation Period." Pp. 79-90 in *Famine as a Geographical Phenomenon*, edited by Bruce Currey and Graeme Hugo. Dordrecht, Netherlands: D. Reidel, 1984.

Winsberg, Morton D. "The Rise and Decline of the American Grapefruit Industry on the Isle of Pines, Cuba." *American Journal of Economics and Sociology*, 20, No. 5 (1961): 543-548.

Zelinsky, Wilbur. "The Historical Geography of the Negro Population of Latin America." *Journal of Negro History*, 34, No. 2 (1949): 153-221.

PART 6

Political Geography

Abreu Garcia, Yadira. "GIS as a Decision Support System for the Evaluation of Environmental Documents, Case Study: Environmental Quality Board of Puerto Rico." Master's Thesis, University of Redlands, Redlands, Calif., 2003.

Aguilar, Adrian G. "Las Bases del Orderamiento Territorial: Algunas Evidencias de la Experiencia Cubana." *Revista Geografica*, 109 (1989): 87-112.

Albuquerque, Klaus de. "Conflicting Claims on the Antigua Coastal Resources: The Case of the McKinnon and Jolly Hill Slat Ponds." Pp. 195-205 in *Caribbean Ecology and Economics*, edited by N.P. Girvon and D. Simmons. Barbados: Caribbean Conservation Association, 1991.

Anderson, Thomas D. "Democracy as a Geopolitical Element Along the Caribbean." *Proceedings, Conference of Latin Americanist Geographers* (1983): 191-199.

Anderson, Thomas D. "Political Power Centers in the Caribbean: Identification and Brief Review." *Ohio Geographer*, 15 (1987): 16-23.

Antonini, Gustavo, Katherine C. Ewel, and John J. Ewel. "Ecological Modeling of a Tropical Watershed: A Guide to Regional Planning." Pp. 51-74 in *Spatial Aspects of Development*, edited by Brian S. Hoyle. London, England: John Wiley, 1974.

Augelli, John P. "The British Virgin Islands: A West Indian Anomaly." *Geographical Review*, 46, No. 1 (1956): 43-58.

Augelli, John P. "Nationalization of Dominican Borderlines." *Geographical Review*, 70, No. 1 (1980): 19-35.

Bahr, Jurgen, and Gunter Mertins. "Regionalpolitik und entwicklung in Kuba 1959-1989." *Geographische Rundschau*, 41, No. 1 (1989): 4-13.

Bahr, Jurgen, and Gunter Mertins. "Politica Regional y Desarrollo Regional en Cuba 1959-1989." *Revista Geografica*, 111 (1990): 5-34.

Barkin, David. "Confronting the Separation of Town and Country in Cuba." *Antipode,* 12, No. 3 (1980): 31-40.

Barsdorf, A. "Kuba: 100% Kubanisch? Anspruch und wirtklichkeit der Entwicklungsstrategie Castros." *Mitteilungen der Osterreichischen Geographischen Gesellschaft*, 135 (1993): 191-202.

Berleant-Schiller, Riva. "Statehood, the Commons, and the Landscape in Barbuda." *Caribbean Geography*, 3, No. 1 (1991): 43-52.

Bleichner, Barry J., III. "An Analysis of Marine Protected Legislation in the Caribbean Lesser Antilles." Master's Thesis, Louisiana State University, Baton Rouge, La., 2008.

Carman, Sheila S. "The Effectiveness of a Small Participatory Water Development Project in the St. Vincent Grenadines, West Indies." Master's Thesis, University of Tennessee, Knoxville, Tenn., 1996.

Chaplowe, Scott. "Havana's Popular Gardens and the Cuban Food Crisis." Master's Thesis, University of California, Los Angeles, Calif., 1996.

Cherni, Judith, and Yohan Hill. "Energy and Policy Providing for Sustainable Rural Livelihoods in Remote Locations-The Case of Cuba." *Geoforum*, 40, No. 4 (2009): 645-654.

Ciccole, William N. "The West Indies: A Political Geography of Federation." Master's Thesis, University of Oklahoma, Norman, Okla., 1965.

Clarke, Colin G. "Political Fragmentation in the Caribbean: The Case of Anguilla." *Canadian Geographer*, 15, No. 1 (1971): 13-29.

Clarke, Colin G. "The Political Ecology of a Town in Trinidad." Pp. 798-801 in *International Geography 1972*, edited by W.P. Adams and F.M. Helleiner. Montreal, Quebec: International Geographical Congress, 1972.

Clarke, Colin G. "Pluralism and Stratification in San Fernando, Trinidad." Pp. 53-70 in *Social Patterns in Cities*, edited by Colin G. Clarke and M.B. Gleave. London, England: Institute of British Geographers, Special Publication, No. 5, 1973.

Clarke, Colin G. "Grenada." Pp. 83-95 in *Politics, Security, and Development in Small States*, edited by Colin G. Clarke and Tony Payne. London, England: Allen and Unwin, 1987.

Clarke, Colin G., ed. *Society and Politics in the Caribbean*. London, England: St. Anthony's Macmillan, 1991.

Clarke, Colin G. "Introduction: Caribbean Decolonisation: New States and Old Societies." Pp. 1-27 in *Society and Politics in the Caribbean*, edited by Colin G. Clarke. London, England: St. Anthony's Macmillan, 1991.

Connell, John. *Bermuda: A Failure of Decolonisation*. Leeds, England: University of Leeds, Department of Geography, Working Paper, No. 492, 1987.

Connell, John. "Island Microstates: Development, Autonomy, and the Ties that Bind." Pp. 117-150 in *The Development Process in Small Island States*, edited by Douglas G. Lockhard, David Drakakis-Smith, and John Schembri. London, England: Routledge, 1993.

Conway, Dennis, and Paul Lorah. "Environmental Protection Policies in Caribbean Small Islands: Some St. Lucian Examples." *Caribbean Geography*, 6, No. 1 (1995): 16-27.

Cook, Ian, and Michelle Harrison. "Cross Over Food: Re-Materializing Postcolonial Geographies." *Transactions, Institute of British Geographers*, 28, No. 3 (2003): 296-317.

Crabot, C. "La Federation des Antilles britanniques." *Information Geographique*, 23 (1959): 47-57.

Dale, Edmund H. "The West Indies: A Federation in Search of a Capital." *Canadian Geographer*, 5, No. 2 (1961): 44-52.

Dale, Edmund H. "The State-Idea: Missing Prop of the West Indies Federation." *Scottish Geographical Magazine*, 78, No. 3 (1962): 166-176.

Day, Michael J. "Stakeholder Reaction to the Proposed Establishment of the National Park, Cockpit Country, Jamaica." Pp. 121-131 in *Ethnography of Protected Areas: Endangered* Habitats-Endangered *Cul-*

tures, edited by P. Simonic. Ljubljana, Slovenia: Filozofska Fakultera, 2006.

Deckert, Emil. "Politisch-geographische Betrachtungen uber Westindien." *Geographische Zeitschrift*, 3 (1897): 121-137.

Dezert, B. "Les Consequences de l'isolement geopolitique de Cuba." *Acta Geographica*, 104 (1995): 3-14.

Dore, P. "Trinidad's Marine Boundaries." *Hydrographic Journal*, 44 (1987): 37-43.

Eggert, K. "Ein Beitrag zur politischen und okonomischen Geographie Cubas." *Zeitschrift fur den Erdkundeunterricht*, 12 (1960): 220-227.

Elliot, S.R., and Ian Lee. "Where the Power Is Wielded." *Geographical Magazine*, 53 (1983): 843-846.

Emery, Byron E. "Regional Integration and Disintegration in the British Caribbean Area." Ph.D. Dissertation, University of Michigan, Ann Arbor, Mich., 1963.

Ewell, Judith. "The Twelve Mile Territorial Sea: New Focus for Conflict in the Caribbean." *Caribbean Studies*, 18, No. 3-4 (1978): 69-88.

Eyre, L. Alan. "Political Violence and Urban Geography in Kingston, Jamaica." *Geographical Review*, 74, No. 1 (1984): 24-37.

Eyre, L. Alan. "Party Political Violence and the Struggle for Residential Space in Jamaica." Pp. 125-139 in *Competition for Rural and Urban Space in Latin America: Its Consequences for Low-Income Groups*, edited by J.M.G. Kleinpenning. Amsterdam, Netherlands: Netherlands Geographical Studies, Band 25, 1986.

Eyre, L. Alan. "The Caribbean Environment: Trends Towards Degradation and Strategies for Their Reversal." *Caribbean Journal of Education*, 16 (1989): 13-45.

Eyre, L. Alan. "Water Resource Conflict on the North Coast: Who Gets What?" *Jamaican Journal*, 22 (1990): 2-11.

Eyre, L. Alan. "Jamaica's Crisis in Forestry and Watershed Management." *Jamaica Naturalist*, 1 (1991): 27-44.

Foucher, Michael. "Les Antilles, iles de la mediterranee d'amerique: Presentation geopolitique." *Notes et etudes documentaires*, 4753 (1984): 33-44.

Gehrmann, V. "Ressourcenmanagement an Grenzen. Ein aktuelle Beispield von der Insel Hispaniola." Pp. 103-110 in *Tagungsband*

Peace Pakrs: Ressourcemanagement an Grenzen, edited by Ludwig Ellenburg. Berlin, Germany: Berliner Geographische Arbeiten, Volume 106, 2006.

Gilbert, Leah. "An Evaluation of Environmental Impact Assessment in Barbados." Master's Thesis, Carleton University, Ottawa, Ontario, 2006.

Glassner, Martin I. "The Foreign Relations of Jamaica and Trinidad and Tobago, 1960-1965." *Caribbean Studies*, 10 (1970): 116-154.

Glassner, Martin I. "CARICOM: A Community in Trouble." *Focus*, 29, No. 2 (1978): 13-16.

Glassner, Martin I. "Management of Marine Resources as a Binding Force in the Eastern Caribbean." *Ocean and Shoreline Management*, 20, No. 1 (1993): 63-88.

Glusa, R. "Zur politischen Geographie Westindiens." Ph.D. Dissertation, Westfalische Wilhelms-Universitat Munster, Munster, Germany, 1962.

Guillermo-Aguilar, Andrian. "Las Bases del Ordenamiendo Territorial: Alguns Evidencias de la Experiencia Cubana." *Revista Geografica*, 109 (1989): 87-112.

Hall, Derek R. "External Relations and Current Development Patterns in Cuba." *Geography*, 66 (1981): 237-240.

Hall, Derek R. "Town and Country Planning in Cuba." *Town and Country Planning*, 50, No. 3 (1981): 81-83.

Hasdenteufel, P. *Natur schutz und Schutzgebiete auf Kuba. Entwicklung und Management am Beispiel zweier Nationalparks*. Munich, Germany: Munchener Geographische Abhandlungen, Volume A56, 2007.

Heather, Nicol. "U.S. Hegemony in the Twenty-First Century: Cuba's Place in the Regionalizing Geopolitics of North America and Caribbean Countries." *Journal of Borderland Studies*, 23, No. 1 (2008): 31-52.

Holcombe, Elizabeth, and Malcolm Anderson. "Tackling Landslide Risk: Helping Land Use Policy to Reflect Unplanned Housing Realities in the Eastern Caribbean." *Land Use Policy*, 27, No. 3 (2010): 798-800.

Hughey, Erin. "A Longitudinal Study: The Impact of a Comprehensive Emergency Management System on Disaster Responses in the

Commonwealth of the Bahamas." Ph.D. Dissertation, University of South Florida, Tampa, Fla., 2008.

Jackiewicz, Edward L. "Bowling for Dollars: Economic Conflicts and Challenges in Contemporary Cuba." *Yearbook, Association of Pacific Coast Geographers*, 64 (2002): 98-111.

Jackiewicz, Edward L., and Todd Bolster. "The Working World of the Palador: The Production of Contradictory Space During Cuba's Period of Fragmentation." *Professional Geographer*, 55, No. 3 (2003): 372-382.

James, Preston E. "Trouble in the Caribbean." *Geographical Perspectives*, 32 (1973): 13-22.

Johnston, R.J., and Michael Morrissey. "The Source of Electoral Bias in Jamaican Election Results, 1967-1980." *Caribbean Geography*, 1, No. 3 (1984): 176-185.

Karol, K.S. "Cuba esta sola: Das Insel-Haerlandeiner Neuen Lateinamerikanischen Hemispher." *Zeitschrift fur Geopolitik*, 39, No. 2 (1968): 82-85.

Kelly, Charles. "Use of Disaster Data in the Planning of Development Projects: Vulnerability Assessment for the Dominican Republic." Master's Thesis, Rutgers University, New Brunswick, N.J., 1987.

Klak, Thomas, ed. *Globalization and Neoliberalism: The Caribbean Context.* Lanham, Md.: Rowman & Littlefield, 1997.

Klak, Thomas. "Development Policy Drift in Central America and the Caribbean." *Singapore Journal of Tropical Geography*, 30, No. 1 (2009): 18-23.

Levine, B.B. "Geopolitical and Cultural Competition in the Caribbean: An Introduction: Cuba Versus the United States." Pp. 1-17 in *The New Cuban Presence in the Caribbean*, edited by B.B. Levine. Boulder, Colo.: Westview, 1983.

Levo, Edith M. "The Federation of the British West Indies: Developments and Prospects." Master's Thesis, Clark University, Worcester, Mass., 1957.

Lopez-Marrero, Tania. "An Integrative Approach to Study and Promote Natural Hazards Adaptive Capacity: A Case Study of Two Flood-Prone Communities in Puerto Rico." *Geographical Journal*, 176, No. 2 (2010): 150-163.

Lowenthal, David. "The West Indies Chooses a Capital." *Geographical Review*, 48, No. 3 (1958): 331-364.

Lowenthal, David. "Two Federations." *Social and Economic Studies*, 6 (1967): 85-96.

Lowenthal, David. "Black Power in the Caribbean Context." *Economic Geography*, 48 (1972): 116-134.

Lowenthal, David, and Colin G. Clarke. "Slave Breeding in Barbuda: The Past of a Negro Myth." *Annals of the New York Academy of Sciences*, 292 (1977): 510-535.

Lowenthal, David, and Colin G. Clarke. "Island Orphans: Barbuda and the Rest." *Journal of Commonwealth and Comparative Politics*, 18 (1980): 293-307.

Lowenthal, David, and Colin G. Clarke. "The Triumph of the Commons: Barbuda Belongs to All Barbudans Together." Pp. 147-158 in *Caribbean Land and Development Revisited*, edited by J. Besson and Janet H. Momsen. New York, N.Y.: Palgrave Macmillan, 2007.

Lusaka, P.F. "The Dissolution of the West Indies Federation: A Study in Political Geography." Master's Thesis, McGill University, Montreal, Quebec, 1964.

Maguire, Robert E. "Sisyphus Revisited: Grassroots Development and Community Conflict in Haiti." *Caribbean Geography*, 5, No. 2 (1994): 127-135.

Miller, Learie, and David Barker. "Land Policy in Jamaica in the Decade After Agenda 21." Pp. 119-134 in *Caribbean Land and Development Revisited*, edited by J. Besson and Janet H. Momsen. New York, N.Y.: Palgrave Macmillan, 2007.

Miller, Lovette A. "Analyzing the Impact of Participatory-Developed Conservation Policies in the Negril Environmental Protection Area, Western Jamaica." Ph.D. Dissertation, University of Maryland, College Park, Md., 2007.

Mings, Robert C. "The Role of the Commonwealth Government in the Growth and Development of the Puerto Rico Tourist Industry." Ph.D. Dissertation, Ohio State University, Columbus, Ohio, 1966.

Momsen, Janet H. "Caribbean Conflict: Cold War in the Sun." *Political Geography Quarterly*, 3, No. 2 (1984): 145-151.

Momsen, Janet H. "Canada-Caribbean Relations: Where is the Special Relationship?" *Political Geography Quarterly*, 11, No. 5 (1992): 501-513.

Mujica-Ortiz, Banery, and Michael J. Day. "Karst Conservation and Protected Areas in Northern Puerto Rico." *Caribbean Geography*, 12, No. 1 (2001): 11-23.

Mullings, Beverley. "Neoliberalization, Social Reproduction, and the Limits to Labour in Jamaica." *Singapore Journal of Tropical Geography*, 30, No. 2 (2009): 174-188.

Newstead, Clare. "Scaling Caribbean (in)dependence." *Geoforum*, 36, No. 1 (2005): 45-58.

Newstead, Clare. "Regional Governmentality: Neoliberalization and the Caribbean Community Single Market and Economy." *Singapore Journal of Tropical Geography*, 30, No. 2 (2009): 158-173.

Nicolas, Thierry. "La circulation comme facteur d'integration nationale et d'hypo-insularitie: Le cas des Antilles francais." *Les Cahiers d'Outre Mer*, 54, No. 216 (2001): 397-416.

O'Keefe, Philip, John Soussen, and Paul Susman. "A Sad Note for Grenada." *Political Geography Quarterly*, 3, No. 2 (1984): 152-160.

Olwig, K.F. "Caribbean Place Identity: From Family Land to Region and Beyond." *Identities*, 5 (1999): 435-467.

Oxtoby, F.E. "The Spatial Impact of Political Processes on the Industrial Structure of the United States Virgin Islands." Master's Thesis, Clark University, Worcester, Mass., 1968.

Oxtoby, F.E. "The Role of Political Factors in the Virgin Islands Watch Industry." *Geographical Review*, 60, No. 4 (1970): 463-474.

Paddington, Luke. "An Appraisal of Environmental Management in Trinidad and Tobago." Master's Thesis, McGill University, Montreal, Quebec, 1999.

Patterson, Karen A. "Political Ecology and Integrated Pest Management in Jamaica." Master's Thesis, Virginia Polytechnic Institute and State University, Blacksburg, Va., 1996.

Pentelaw, Laurel. "Climate Policy and International Tourism Arrivals to the Caribbean Region." Master's Thesis, University of Waterloo, Waterloo, Ontario, 2009.

Platt, Raye R. "A Note on Political Sovereignty and Administration in the Caribbean." *Geographical Review*, 16, No. 4 (1926): 623-637.

Poncet, J. "Les transformation de l'espace de Cuba et al nouvelle geographie administrative." *L'Espace Geographique*, 5 (1976): 255-262.

Possekel, Anja K. "Complex Resource Management on Small Caribbean Islands: The Potential of Scenarios in Planning Approaches." Pp. 30-43 in *Land, Sea, and Human Effort in the Caribbean*, edited by B. Ratter and W. Sahr. Hamburg, Germany: Beitrage zur Geographischen Regionalforschung in Lateinamerika, Volume 10, 1997.

Potter, Robert B. "Basic Need and Development in the Small Island States of the Eastern Caribbean." Pp. 92-116 in *The Development Process in Small Island States*, edited by Douglas G. Lockhart, D. Drakakis-Smith, and J. Schembri. London, England: Routledge, 1993.

Potter, Robert B., and Jonathan Pugh. "Planning Without Plans and the Neo-Liberal State: The Case of St. Lucia, West Indies." *Third World Planning Review*, 23 (2001): 323-340.

Potter, Robert B., and Mark Watson. "Current Housing Policy Issues in Barbados: With Particular Reference to Vacant Sub-Divisions." *Third World Planning Review*, 21 (1999): 259-282.

Pugh, Jonathan. "On Communicative Action and Power: The National Commission on Sustainable Development in Barbados." *Caribbean Geography*, 12, No. 1 (2001): 1-10.

Pugh, Jonathan. "Environmental Planning in Barbados: A Confident State, Isolated Environmental Movements, and Anxious Development Consultants." *Southeastern Geographer*, 45, No. 2 (2005): 192-205.

Pugh, Jonathan. "The Disciplinary Effects of Communicative Planning in Soufriere, St. Lucia: The Relative Roles of Governmentality, Hegemony, and Space-Time Politics." *Transactions, Institute of British Geographers*, 30, No. 3 (2005): 307-322.

Pugh, Jonathan. "The Participation Paradox: Stories from St. Lucia." Pp. 95-106 in *Caribbean Land and Development Revisited*, edited by J. Besson and Janet H. Momsen. New York, N.Y.: Palgrave Macmillan, 2007.

Pugh, Jonathan, and Janet H. Momsen. *Environmental Planning in the Caribbean*. Burlington, Vt.: Ashgate, 2006.

Pugh, Jonathan, and Robert B. Potter. "Rolling Back the State and Physical Development Planning: The Case of Barbados." *Singapore Journal of Tropical Geography*, 21, No. 2 (2000): 183-199.

Pugh, Jonathan, and Robert B. Potter. *Participatory Planning in the Caribbean: Lessons from Practice*. Burlington, Vt.: Ashgate, 2003.

Pugh, Jonathan, and Robert B. Potter. "Participatory Planning in the Caribbean: Some Key Themes." Pp. 1-22 in *Participatory Planning in the Caribbean: Lessons from Practice*, edited by Jonathan Pugh and Robert B. Potter. Burlington, Vt.: Ashgate, 2003.

Ratter, B. "Resource Management Changes in the Caribbean: The Eco-Eco-Approach." Pp. 17-30 in *Land, Sea, and Human Effort in the Caribbean*, edited by B. Ratter and W. Sahr. Hamburg, Germany: Beitrage zur Geographischen Regionalforschung in Lateinamerika, Volume 10, 1997.

Ratter, B., and Gerhard Sandner, eds. *Territorialkonflikte im karibischen Meeresraum-Interessenhintergrunde, Stilformen, und Losungsansatze*. Hamburg, Germany: Universitat Hamburg, Beitrage zur Geographischen Regionalforschung in Lateinamerika, Volume 9, 1993.

Reid-Henry, Simon. "The Contested Spaces of Cuban Development: Post-Socialism, Post-Colonialism, and the Geography of Transition." *Geoforum*, 38, No. 3 (2007): 445-455.

Revert, E. "Geographie politique du monde Caraibe." *Annales de Geographie*, 6o (1950): 34-47.

Riddell, J.B. "The Face of Neo-Liberalism in the Third World: Landscapes of Coping in Trinidad and Tobago." *Canadian Journal of Development Studies*, 24, No. 4 (2003): 593-615.

Riddell, J.B. "A Tale of Contestation, Disciples, and Damned: The Lessons of the Spread of Globalization into Trinidad and Tobago." *Environment and Planning A*, 34, No. 4 (2003): 659-678.

Robertson, Richard E., W.B. Ambeh, and Lloyd Lynch. "Strategic Planning for Volcanic Emergencies in the Commonwealth Eastern Caribbean." *Caribbean Geography*, 6, No. 2 (1995): 77-96.

Roucek, Joseph S. "Cuba in Its Geopolitical Setting." *Contemporary Review*, 198 (1960): 663-667.

Roucek, Joseph S. "Haiti in Geopolitics." *Contemporary Review*, 204 (1963): 21-29.

Roucek, Joseph S. "The Dominican Republic in Geopolitics." *Contemporary Review*, 206 (1965): 289-294; 207 (1965): 12-21.

Royle, Stephen A. "The Small Island as Colony." *Caribbean Geography*, 3, No. 4 (1992): 261-269.

Sagawe, Thorsten. "Caudillismo und Stadterntwicklung: Das Beispiel von Santo Domingo." *Geographische Zeitschrift*, 80 (1992): 51-64.

Saint-Vil, J. "Villes et bourgs de Saint-Dominque au XVIIIe siecle." *Les Cahiers d'Outre Mer,* 31 (1978): 251-270.

Sandner, Gerhard. "Politisch-Geographische Raumstrukturen und Geopolitik im Karibischen Raum." *Geographische Zeitschrift*, 69, No. 1 (1981): 34-56.

Santana, Deborah B. *Kicking off the Bootstraps: Environments, Development, and Community Power in Puerto Rico*. Tucson, Ariz.: University of Arizona Press, 1996.

Santana, Deborah B. "Geographers, Colonialism, and Development Strategies: The Case of Puerto Rico." *Urban Geography*, 17, No. 5 (1996): 456-474.

Schafer, Otto. "Geopolitik des Karibischen Raumes." *Geographischer Anzeiger*, 45, No. 1-4 (1944): 2-7.

Serbin, Andres. *Caribbean Geopolitics: Toward Security Through Peace?* Boulder, Colo.: Lynne Rienner, 1990.

Simey, T.S. "A New Capital for the British West Indies." *Town Planning Review*, 28, No. 1 (1957): 63-70.

Skelton, Tracey, and Susan P. Mains. "Intersections of Neoliberalization, Mobilities, and Development in the Caribbean: An Introduction." *Singapore Journal of Tropical Geography*, 30, No. 2 (2009): 151-157.

Sletto, Bjorn. "Landscape of Resistance: Reading, Producing, and Representing the Nariva Swamp, Trinidad." Master's Thesis, University of Kansas, Manhattan, Kans., 1999.

Sletto, Bjorn. "Producing Space(s), Representing Landscapes: Maps and Resource Conflicts in Trinidad." *Cultural Geographies*, 9, No. 4 (2002): 389-420.

Sletto, Bjorn. "A Swamp and Its Subjects: Conservation Politics, Surveillance, and Resistance in Trinidad, the West Indies." *Geoforum*, 36, No. 1 (2005): 77-93.

Stodder, Joseph H., and Kevin F. McCarthy. *Profiles of the Caribbean Basin in 1960-1980: Changing Geopolitical and Geostrategic Dimensions*. Santa Monica, Calif.: Rand Note No. N-2058-AF, 1983.

Trist, Carolyn. "Environmental Security, Conservation, and Resources Entitlements in the Eastern Caribbean." Ph.D. Dissertation, University of California, Berkeley, Calif., 2000.

Turner, Scott D. "Evaluation of Vervot Monkey Trapping on the Island of Barbados: Using GIS and Adaptive Management to Manage Conflicting Resources." Master's Thesis, Central Washington University, Ellensburg, Wash., 2001.

Ungefehr, F. "Freeport/Grand Bahama: ein privates Entwicklungsprojekt in den Netzen der bahamischen Innenpolitik." *Die Erde*, 119, No. 2 (1988): 99-110.

Watts, David J. "Environmental Degradation, the Water Resource, and Sustainable Development in the Eastern Caribbean." *Caribbean Geography*, 6, No. 1 (1995): 2-15.

Whittlesey, D.S. "Geographic Factors in the Relations of the United States and Cuba." *Geographical Review*, 12 (1922): 241-266.

Woodward, R.L. "Crisis in the Caribbean: Trouble in Uncle Sam's Backyard." *Geographical Magazine*, 53 (1981): 838-842.

Part 7

Urban Geography

Albuquerque, Klaus de, Wesley Van Riel, and J. Mark Taylor. "Uncontrolled Urbanization in the Developing World: A Jamaican Case Study." *Journal of Developing Areas*, 14, No. 3 (1980): 349-360.

Ammerl, T. *Aktuelle Stadt und Landschaftsokologische Probleme in Havanna und Losungsanstatze durch Staatliche Raumordnung, Umwelt politik bzw*. Munich, Germany: Munchen Geographische Abhandlungen, Volume A57, 2007.

Ammerl, T. "Diewahrnehmug stadtkologischer Probleme durch die Bevolkerung Havannas." Pp. 83-98 in *Stadtwelten in den Americas*, edited by Eberhard Rothfuss and Werner Gamerith. Passau, Germany: Passauer Schriften zur Geographie, Volume 23, 2007.

Ammerl, T. "Aktuelle okologische Probleme in Havanna (Kuba)." *Geographische Rundschau*, 59, No. 1 (2007): 58-64.

Appleton, Nathan. "Santo Domingo City." *Journal of the American Geographical Society of New York*, 23 (1891): 533-565.

Bailey, Wilma. "Child Morbidity in the Kingston Metropolitan Area, Jamaica, 1983." *Social Science and Medicine*, 26, No. 11 (1988): 1117-1124.

Bailey, Wilma, and D.R. Phillips. "Spatial Patterns of Use of Health Services in the Kingston Metropolitan Area, Jamaica." *Social Science and Medicine*, 30, No. 1 (1989): 1-12.

Barsotti, Natasha. "Social Processes and Environmental Factors Contributing to the Development of Steelbands in Port of Spain, Trinidad." Master's Thesis, University of Texas, Austin, Tex., 1996.

Bennett, Helen C. "Kingston, the Capital of Jamaica, as It Was and Is." *Bulletin of the Geographical Society of Philadelphia*, 5 (1907): 73-81.

Borsch, D. "Metroplisierung in Entwicklungslanden unter unterschiedlischen soziookonomischen Bedingungen. Das Beispiel Havana. Eine Unterrichsreihe fur den Kursunterriht." *Geographie und Schule*, 11, No. 61 (1989): 25-35.

Boswell, Thomas D., and J. Briggs. "The Bahamas." Pp. 252-284 in *Urbanization, Planning, and Development in the Caribbean*, edited by Robert B. Potter. London, England: Mansell, 1989.

Boswell, Thomas D. "Characteristics and Processes of Urbanization in the Caribbean." *Yearbook, Conference of Latin Americanist Geographers*, 17-18 (1990): 67-90.

Chaline, C. "La Havane: Urbanisme de rapture ou de rattragage?" *Annales de Geographie*, 96 (1987): 171-185.

Chamberlin, Thomas W. "Phases of the Urban and Industrial Geography of Havana, Cuba." Ph.D. Dissertation, Clark University, Worcester, Mass., 1946.

Chaplowe, Scott. "Havana's Popular Gardens and the Cuban Food Crisis." Master's Thesis, University of California, Los Angeles, Calif., 1996.

Clarke, Colin G. "Population Pressure in Kingston, Jamaica: A Study of Unemployment and Overcrowding." *Transactions, Institute of British Geographers*, 38 (1966): 165-182.

Clarke, Colin G. "Aspects of the Urban Geography of Kingston, Jamaica." Ph.D. Dissertation, University of Oxford, Oxford, England, 1967.

Clarke, Colin G. "An Overcrowded Metropolis: Kingston, Jamaica." Pp. 305-328 in *Geography and a Crowding World*, edited by Wilbur Zelinsky, Leszak A. Kosinski, and R.M. Prothero. New York, N.Y.: Oxford University Press, 1970.

Clarke, Colin G. "The Development and Redevelopment of the Waterfront in Kingston, Jamaica." *Geography*, 56 (1971): 237-240.

Clarke, Colin G. "Pluralism and Stratification in San Fernando, Trinidad." Pp. 53-70 in *Social Patterns in Cities*, edited by B.D. Clark and M.B. Gleave. London, England: Institute of British Geographers, Special Publication, No. 5, 1973.

Clarke, Colin G. "Urbanization in the Caribbean." *Geography*, 59 (1974): 223-232.

Clarke, Colin G. *Kingston, Jamaica: Urban Growth and Social Change, 1692-1962*. Berkeley, Calif.: University of California Press, 1974.

Clarke, Colin G. "Ecological Aspects of Population Growth in Kingston, Jamaica." Pp. 42-55 in *Geographical Analysis for Development in Latin America and the Caribbean*, edited by R.P. Momsen. Chapel Hill, N.C.: Conference of Latin Americanist Geographers Publications, 1975.

Clarke, Colin G. "Dependency and Marginality in Kingston, Jamaica." *Journal of Geography*, 82, No. 5 (1983): 227-235.

Clarke, Colin G. *Decolonizing the Colonial City: Urbanization and Stratification in Kingston, Jamaica*. New York, N.Y.: Oxford University Press, 2006.

Clarke, Colin G. *Kingston: Urban Development and Social Change*. Kingston, Jamaica: Ian Randle, 2006.

Clarke, Colin G. "From Slum to Ghetto: Social Deprivation in Kingston, Jamaica." *International Development Planning Review*, 28, No. 1 (2006): 1-34.

Clarke, Colin G. "From Slum to Ghetto: Multiple Deprivation in Kingston, Jamaica." Pp. 261-278 in *Global Change and Caribbean Vulnerability: Environment, Economy, and Society at Risk*, edited by Duncan McGregor, David Dodman, and David Barker. Mona, Jamaica: University of West Indies Press, 2009.

Clarke, Colin G., and David Howard. "Race and Religious Pluralism in Kingston, Jamaica." *Population, Space, and Place*, 11, No. 2 (2005): 119-136.

Clarke, Colin G., and David Howard. "Contradictory Socio-Economic Consequences of Structural Adjustment in Kingston, Jamaica." *Geographical Journal*, 172, No. 2 (2006): 106-129.

Cocq, Karen. "Change, Continuity, and Contradiction in the Cuban Waterscape: Privatization and the Case of Aguas De La Habana." Master's Thesis, Queen's University, Kingston, Ontario, 2006.

Collins, William. "Residential Area Change and Residential Relocation in Port of Spain, Trinidad." Ph.D. Dissertation, University of Texas, Austin, Tex., 1977.

Colon, Christobal. "Discrimination des utilisations du sol en milieu urbain par teledetection aerospatiale a Santiago de Los Caballeros, Republique Dominique." Master's Thesis, University of Sherbrooke, Sherbrooke, Quebec, 1993.

Colontonio, A., and Robert B. Potter. "Havana." *Cities*, 23, No. 1 (2006): 63-78.

Colontonio, A., and Robert B. Potter. "The Rise of Urban Tourism in Havanna Since 1989." *Geography*, 91, No. 1 (2006): 23-33.

Conway, Dennis. "Residential Area Change and Residential Relocations in Port-of-Spain, Trinidad." Ph.D. Dissertation, University of Texas, Austin, Tex., 1977.

Cumpter, Gloria. "New Pattern for Kingston." *Geographical Magazine*, 60, No. 7 (1967): 588-598.

Curtis, James R. "Havana's Parque Coppelia: A Microcosm of Public Space Traditions in Socialist Cuba." *Places*, 8, No. 3 (1993): 62-67.

DeGraaff, Jan. "Evaluating the Sustainable Development of the Kingston Watersheds in Jamaica, 1980-1993." *Caribbean Geography*, 8, No. 1 (1997): 46-56.

Desse, Michel. "Les nouvelles formes de polarisation urbaine en Guadeloupe, Martinique, et Reunion." *Cahiers de Geographie du Quebec*, 42, No. 116 (1998): 223-248.

Desse, Michel. "Hierarchies et polarisations urbaines recentes en Guadeloupe, Martinique, et Reunion." *Les Cahiers d'Outre Mer*, 51 (1998): 65-88.

Desse, Michel. "Les villes de la Caraibe et la Mer." *Etudes Caribeennes*, 2 (2005): 51-67.

Devauges, R. "Une capitale antillaise: Port-au-Prince (Haiti)." *Les Cahiers d'Outre Mer*, 7 (1954): 105-136.

Diaz-Garayua, Jose R. "Neighborhood Characteristics and Housing Values within the San Juan MSA, Puerto Rico." *Southeastern Geographer*, 49, No. 4 (2009): 376-393.

Dodman, David. "Shooting in the City: An Autophotographic Exploration of the Urban Environment in Kingston, Jamaica." *Area*, 35, No. 3 (2003): 293-304.

Dodman, David. "Community Perspectives on Urban Environmental Problems in Kingston, Jamaica." *Social and Economic Studies*, 53, No. 3 (2004): 31-59.

Dodman, David. "Post-Independence Optimism and the Legacy of Waterfront Redevelopment in Kingston, Jamaica." *Cities*, 24, No. 4 (2007): 273-284.

Dodman, David. "Developers in the Public Interest? The Role of Urban Development Corporations in the Anglophone Caribbean." *Geographical Journal*, 174, No. 1 (2007): 30-44.

Donner, W. "Port-au-Prince und seine metropolitan-Zone: Eine wirtschaftsgeographische Skizze der Hauptstadt Haitis." *Zeitschrift fur Wirtschaftsgeographie*, 21, No. 7 (1977): 209-214.

Doran, Michael F., and Renee A. Landis. "Origin and Persistence of an Inner-City Slum in Nassau." *Geographical Review*, 70, No. 2 (1980): 182-193.

Douzant-Rosenfeld, Denise. "Aspects de l'approvisionnement alimentaire d'une capitale du Tiers Monde: Santo Domingo, Republique Dominicaine." *L'Espace Geographique*, 16 (1987): 101-114.

Dyer, Donald R. "Urbanism in Cuba." *Geographical Review*, 47, No. 2 (1957): 224-233.

Edward, Carol J. "The Older Generation: An Ecological Study of Old Age in Havana." Ph.D. Dissertation, Syracuse University, Syracuse, N.Y., 1979.

Eyre, L. Alan. "The Shantytowns of Montego Bay, Jamaica." *Geographical Review*, 62, No. 3 (1972): 394-413.

Eyre, L. Alan. "The Shantytowns of Montego Bay, Jamaica." *Ekistics*, 36 (1973): 132-138.

Eyre, L. Alan. "Quasi-Urban Melange Settlement: Cases for St. Catherine and St. James, Jamaica." *Geographical Review*, 69, No. 1 (1979): 95-100.

Eyre, L. Alan. "The Internal Dynamics of Shanty Towns in Jamaica." *Caribbean Geography*, 1, No. 4 (1984): 256-270.

Eyre, L. Alan. "The Effects of Political Terrorism on the Residential Location of the Poor in the Kingston Urban Region, Jamaica, West Indies." *Urban Geography*, 7, No. 3 (1986): 227-242.

Garcia-Ellin, Juan C. "Gated Communities and Housing Projects: The Control of Public Space in San Juan." *Southeastern Geographer*, 49, No. 4 (2009): 354-375.

Gergel, Thomas J. "Monitoring Squatter Settlements in the Developing World: A Case Study in San Juan, Puerto Rico." Master's Thesis, University of South Carolina, Columbia, S.C., 1987.

Godard, H. "Port-au-Prince (Haiti): Les quartiers et les mutations recentes du tissu urbain." *Les Cahiers d'Outre Mer*, 38 (1985): 5-24.

Goudet, Francoise. *Le quartier de l'assainissement a Pointe-a-Pitre, Guadeloupe: Contribution a l'etudes des phenomenes de croissance et de reovations urbaines en milieu tropicale*. Talence, France: Travaux et documents de geographie tropicale, No. 19, 1973.

Hall, Derek R. "Cuba." Pp. 77-113 in *Urbanization and Development in the Caribbean*, edited by Robert B. Potter. London, England: Mansell, 1989.

Hanson, Gerald T. "Shantytown Stage Development: The Case of Kingston, Jamaica." Ph.D. Dissertation, Louisiana State University, Baton Rouge, La., 1975.

Holland, Jeremy. "Global Process, Local Change: Adjustment in Urban Jamaica." *Caribbean Geography*, 4, No. 2 (1993): 89-101.

Howard, David. "Urban Violence, Crime, and the Threat to Democratic Security in the Dominican Republic." Pp. 298-316 in *Global Change and Caribbean Vulnerability: Environment, Economy, and Society at Risk*, edited by Duncan McGregor, David Dodman, and David Barker. Mona, Jamaica: University of West Indies Press, 2009.

Hubbard, Raymond. *The Spatial Pattern and Market Areas of Urban Settlement in Jamaica*. Mona, Jamaica: University of the West Indies, Department of Geography, Paper No. 1, 1970.

Hudson, Brian J. "Urbanization and Planning in the West Indies." *Caribbean Quarterly*, 26, No. 3 (1980): 1-17.

Hudson, Brian J. "The Commonwealth Eastern Caribbean." Pp. 181-211 in *Urbanization, Planning, and Development in the Caribbean*, edited by Robert B. Potter. London, England: Mansell, 1989.

Huetz de Lemps, C. "La croissance de Kingston (Jamaique)." *Les Cahiers d'Outre Mer*, 32 (1979): 103-107.

Ishmael, Len. "Urbanization Dynamics in the Eastern Caribbean: Focus on the Windward Islands." *Cities*, 8, No. 3 (1991): 174-192.

Klak, Thomas. "Havana and Kingston: Mass Media Images and Empirical Observations of Two Caribbean Cities in Crisis." *Urban Geography*, 15, No. 4 (1994): 318-344.

Kush, Melanie J. "Port of Spain: A Sense of Place." Master's Thesis, San Diego State University, San Diego, Calif., 1984.

Lacina, J. et al. "Environment of the Municipium Los Palacios in Cuba." Pp. 211-226 in *Analysis and Synthesis of Geographic Systems*. Brno, Czechoslovakia: Czechoslovak Academy of Sciences, 1992.

Lloyd-Evans, Sally, and Robert B. Potter. "Environmental Impacts of Urban Development and the Urban Informal Sector in the Caribbean." Pp. 245-260 in *Land Degradation in the Tropics*, edited by M.J. Eden and J.T. Parry. London, England: Pinder, 1996.

Lloyd-Evans, Sally, and Robert B. Potter. "Environmental Impacts of Urban Development and the Urban Informal Sector in the Caribbean." Pp. 153-171 in *The Urban Caribbean in an Era of Global Change*, edited by Robert B. Potter. Aldershot, England: Ashgate, 2000.

Mains, Susan P. "Monumentally Caribbean: Borders, Bodies, and Redemptive City Spaces." *Small Axe*, 16 (2004): 179-198.

Martouzet, D. "Urban and Spatial Consequences of Affective Relation to the City: The Case of Fort de France, Martinique (in French)." *Annales de Geographie*, 111, No. 1 (2002): 73-85.

Mertins, Gunter. "Stadtetourismus in Havanna (Kuba)." *Geographische Rundschau*, 55, No. 5 (2003): 20-25.

Muraya, Petronella W.K. "Community-Based Organizations (CBO) Performance in the Shelter Sector: A Study of La Pine and Lebron, Santo Domingo, Dominican Republic." *Urban Geography*, 27, No. 2 (2006): 152-173.

Newling, Bruce E. "The Growth and Spatial Structure of Kingston, Jamaica." Ph.D. Dissertation, Northwestern University, Evanston, Ill., 1962.

Nickel, A. "Die Alstadt von La Habana: Wohnsituation und Konzepte der Alstadterneurung." *Geographische Rundschau*, 41, No. 1 (1989): 14-21.

Nickel, A. "El Casco Historico de la Habana: La Situacion de vivienda y los conceptos de renovacion urbana en la Habana. *Revista Geographica*, 112 (1990): 75-90.

Norton, Ann. *Shanties and Skyscrapers: Growth and Structure of Modern Kingston*. Mona, Jamaica: University of the West Indies Press, Working Paper No. 13, 1978. 116 pp.

Paquette, Romain. "Une cite planifee et une cite spontanee. (Fort-de-France, Martinique)." *Cahiers de Geographie de Quebec*, 13, No. 29 (1969):169-186.

Pichler, A. "La Habana Bruja. Rituelle Geographie in der Alstadt von Havanna." Pp. 115-128 in *Lateinamerika im Umbruch: Geistige Stromungen im Globalisierungsstress: Beitrages zu einer Ring vorlesung im Winters mester 2001-02 on den Univesitaten Innsbruck und Wien*. Innsbruck, Austria: Innsbrucker Geographische Studien, Band 32, 2001.

Plantin, Corinne. "La diffusion des cultures urbaines etat-uniennes dans l'agglomeration de Fort-de-France." *Les Cahiers d'Outre Mer*, 62 (2009): 241-255.

Ponti, A. "Una rivoluzione a tutela di una citta el caso de Avana." *Bolletino della Societa Geografica Italiana*, Serie 12, 4, No. 3 (1999): 491-510.

Ponton, Marin. "Diagnostic et perspectives d'un systeme de gestion des dechets domestiques urbains: Le cas de Santiago do los Caballeros en Republique Dominicaine." Master's Thesis, University of Sherbrooke, Sherbrooke, Quebec, 1995.

Potter, Robert B. "Urban Development, Planning, and Demographic Change, 1970-1980, in Barbados." *Caribbean Geography*, 1, No. 1 (1983): 3-12.

Potter, Robert B. "Spatial Perceptions and Public Involvement in Third World Urban Planning: The Example of Barbados." *Singapore Journal of Tropical Geography*, 5, No. 1 (1984): 30-44.

Potter, Robert B. *Urbanization, Planning, and Development in the Caribbean*. London, England: Mansell, 1989.

Potter, Robert B. "Urbanization, Planning, and Development in the Caribbean: An Introduction." Pp. 1-20 in *Urbanization, Planning, and Development in the Caribbean*, edited by Robert B. Potter. London, England: Mansell, 1989.

Potter, Robert B. "Urbanization in the Caribbean and Trends of Global Convergence-Divergence." *Geographical Journal*, 159, No. 1 (1993): 1-21.

Potter, Robert B. "Urbanization and Development in the Caribbean." *Geography*, 80, No. 4 (1995): 334-341.

Potter, Robert B. "The Port of Spain Urban Corridor." *Caribbean Geography*, 7, No. 1 (1996): 75-80.

Potter, Robert B. "From Plantopolis to Mini-Metropolis in the Eastern Caribbean: Reflections on Urban Sustainability." Pp. 51-68 in *Resource Sustainability and Caribbean Development*, edited by Duncan McGregor et al. Mono, Jamaica: University of West Indies Press, 1998.

Potter, Robert B. *The Urban Caribbean in an Era of Global Change*. Brookfield, Vt.: Ashgate, 2000.

Potter, Robert B. "Urban Castries, St. Lucia Revisited: Global Forces and Local Responses." *Geography*, 86, No. 4 (2001): 329-336.

Potter, Robert B., and Muriel L. Hunte. "Recent Developments in Planning the Settlement Hierarchy of Barbados: Implications Concerning the Debate on Urban Primacy." *Geoforum*, 10 (1979): 355-362.

Potter, Robert B., and E. McAslon. "Urban Poverty and the Urban Development Commission in Barbados, West Indies." *Geography*, 85 (2000): 263-267.

Potter, Robert B., and Mark Wilson. "Barbados." Pp. 114-139 in *Urbanization, Planning, and Development in the Caribbean*, edited by Robert B. Potter. London, England: Mansell, 1989.

Reece, Andrew G. "Colonial Dualism in the Commonwealth Caribbean: A Study of Kingston, Jamaica." Master's Thesis, University of North Carolina, Chapel Hill, N.C., 1994.

Riddell, J.B. "On Mobility in the Southern City's Suburbs: Residential Relocation in Trinidad and Tobago." *Applied Geographical Studies*, 1, No. 2 (1997): 115-128.

Sagawe, Thorsten. "Drogenphanomene als stadtgeographische Fragestellung am Beispiel von Santo Domingo." *Geographische Zeitschrift*, 77, No. 3 (1989): 168-179.

Sagawe, Thorsten. "Die Altstadt von Santo Domingo." *Geographische Rundschau*, 44, No. 9 (1992): 536-542.

Sahr, W. "Micro-Metropolis in the Eastern Caribbean: The Example of St. Lucia." Pp. 69-86 in *Resource Sustainability and Caribbean Development*, edited by Duncan McGregor et al. Mono, Jamaica: University of West Indies Press, 1998

Samson, M.L. "The Netherlands Antilles." Pp. 161-180 in *Urbanization, Planning, and Development in the Caribbean*, edited by Robert B. Potter. London, England: Mansell, 1989.

Scarpaci, Joseph L. "Reshaping Habana Vieja: Revitalization, Historic Preservation, and Restructuring in the Socialist City." *Urban Geography*, 21, No. 8 (2000): 724-744.

Schneider-Sliwa, Rita. "Exogenous and Endogenous Concepts of Urban Planning: Urban Revitalization in Kingston, Jamaica." *Die Erde*, 122 (1991): 3-22.

Sciarillo, Emily. "Urban Agriculture in Havana: Creating a Sustainable Food Production System." Master's Thesis, George Washington University, Washington, D.C., 2008.

Seepersad, Mary B. "Urban Population Density Gradients in Port of Spain, Trinidad." Master's Thesis, Michigan State University, East Lansing, Mich., 1982.

Semple, Hugh. "Downtown Revitalisation in Kingston, Jamaica: An Assessment." *Caribbean Geography*, 10, No. 2 (1999): 89-102.

Semple, Hugh, and Marina Herbert. "Towards a Model for Revitalizing Older Commercial Strips in Downtown Kingston, Jamaica." *Caribbean Geography*, 12, No. 2 (2001): 55-69.

Senninger, E.J. "A Geographic Analysis of New Providence, Bahamas." Master's Thesis, Illinois State University, Normal, Ill., 1955.

Stetson, George E. "San Juan, Puerto Rico: A Case Study of the Evolution and Functional Role of a Private City." Ph.D. Dissertation, University of North Carolina, Chapel Hill, N.C., 1977.

Strasser S. "Die Zunahme soziairraumlicher Polarisierung in Havanna seit der wirtschaftskrise der 1990er Jahre." *Mitteilungen der Ostterreichischen Geographischen Gesellschaft*, 148 (2006): 205-222.

Tata, Robert J., and Steven J. Tata. "Haiti." Pp. 294-312 in *Latin American Urbanization: Historical Profiles of Major Cities*, edited by Gerald M. Greenfield. Westport, Conn.: Greenwood, 1994.

Theodat, J.M. "L'eau dans l'Etat, l'Etat nan dlo: Dilution des responsabilites publiques et crise urbaine dans l'aire metropolitaine de Port-au-Prince." *Bulletin de l'Association de Geographes Francais*, 87, No. 3 (2010): 336-350.

Thomas, Gerald A. "The Caribbean Urban Spectacle: Race, Class, and Representation in the Social Production of Space." Ph.D. Dissertation, University of Kentucky, Lexington, Ky., 1993.

Thomas, Karen. "Post-Independence Transformation of Secondary Urban Places in Jamaica." *Caribbean Geography*, 13, No. 2 (2003): 114-129.

Thraves, Bernard D. "Urbanization and Planning Issues in Montego Bay, Jamaica." Pp. 225-244 in *Resources, Planning, and Environmental Management in a Changing Caribbean*, edited by David Barker and Duncan McGregor. Kingston, Jamaica: University of West Indies Press, 2003.

Tillman, Benjamin F. "Measuring Globalization's Influence on the Cultural Landscape: Spatial Succession in the Plaza of Ponce, Puerto Rico." *Southeastern Geographer*, 49, No. 4 (2009): 340-353.

Treasure, Allison. "Analysis of Urban Consumer Water-Use Behavior Under Rationing Condition in Kingston, Jamaica." Master's Thesis, South Illinois University, Carbondale, Ill., 2001.

Tupper, Howard M. "Urban Morphological Theory and Spatial Differentiation in a Caribbean City: Residential Land Use from 1948 to 1975 in Santo Domingo, the Dominican Republic." Master's Thesis, University of Florida, Gainesville, Fla., 1982.

Weaver, David B. "Model of Urban Tourism for Small Caribbean Islands." *Geographical Review*, 83, No. 2 (1993): 134-140.

Williams, Diana. "Preparing for the Future: Sustainable Development and Tourism in Havana's Old City." Pp. 71-79 in *Land, Sea, and Human Effort in the Caribbean*, edited by B. Ratter and W. Sahr.

Hamburg, Germany: Beitrage zur Geographischen Regionalforschung in Lateinamerika, Volume 10, 1997.

AUTHOR INDEX

www.ingramcontent.com/pod-product-compliance
Ingram Content Group UK Ltd.
Pitfield, Milton Keynes, MK11 3LW, UK
UKHW041123071026
15587UKWH00009B/156

9 780810 883031